$16.95

787

mu

Volume 9

SCIENCE EDUCATION
AND INFORMATION TRANSFER

Science and Technology Education and Future Human Needs

General Editor: JOHN LEWIS

Malvern College, United Kingdom

Related Pergamon Journal

INTERNATIONAL JOURNAL OF EDUCATIONAL DEVELOPMENT*

Editor: PHILIP TAYLOR

Throughout the world educational developments are taking place: developments in literacy, programmes in vocational education, in curriculum and teaching, in the economics of education and in educational administration.

It is the purpose of the *International Journal of Educational Development* to bring these developments to the attention of professionals in the field of education, with particular focus upon issues and problems of concern to those in the Third World. Concrete information, of interest to planners, practitioners and researchers, is presented in the form of articles, case studies and research reports.

*Free specimen copies available on request.

SCIENCE EDUCATION AND INFORMATION TRANSFER

Edited by

C. A. TAYLOR
University College, Cardiff

Published for the

ICSU PRESS

by

PERGAMON PRESS
OXFORD · NEW YORK · BEIJING · FRANKFURT
SÃO PAULO · SYDNEY · TOKYO · TORONTO

U.K.	Pergamon Press, Headington Hill Hall, Oxford OX3 0BW, England
U.S.A.	Pergamon Press, Maxwell House, Fairview Park, Elmsford, New York 10523, U.S.A.
PEOPLE'S REPUBLIC OF CHINA	Pergamon Press, Qianmen Hotel, Beijing, People's Republic of China
FEDERAL REPUBLIC OF GERMANY	Pergamon Press, Hammerweg 6, D-6242 Kronberg, Federal Republic of Germany
BRAZIL	Pergamon Editora, Rua Eça de Queiros, 346, CEP 04011, São Paulo, Brazil
AUSTRALIA	Pergamon Press Australia, P.O. Box 544, Potts Point, N.S.W. 2011, Australia
JAPAN	Pergamon Press, 8th Floor, Matsuoka Central Building, 1-7-1 Nishishinjuku, Shinjuku-ku, Tokyo 160, Japan
CANADA	Pergamon Press Canada, Suite 104, 150 Consumers Road, Willowdale, Ontario M2J 1P9, Canada

Copyright © 1987 ICSU Press

First edition 1987

Library of Congress Cataloging in Publication Data

Science education and information transfer.
(Science and technology education and future human needs; vol 9)
Papers from the Bangalore Conference on Science and Technology Education and Future Human Needs, held in Bangalore, India, Aug. 6–14, 1985, which was organized by the Committee on the Teaching of Science of the International Council of Scientific Unions.
1. Science—Study and teaching—Congresses.
2. Science—Information services—Congresses.
3. Communication in science—Congresses.
4. Communication of technical information—Congresses.
I. Taylor, Charles Alfred. II. Bangalore Conference on Science and Technology Education and Future Human Needs (1985). III. International Council of Scientific Unions. Committee on the Teaching of Science. IV. Title: Science education and information transfer. V. Series.
Q181.A1S33   1987   507'.1   86–25229

British Library Cataloguing in Publication Data

Science education and information transfer.
(Science and technology education and future human needs; vol 9)
1. Technical education.
I. Taylor, C. A. II. Series.
607   T65

ISBN 0-08-033954-9 (Hardcover)
ISBN 0-08-033955-7 (Flexicover)

D
607
SCI

Printed in Great Britain by A. Wheaton & Co. Ltd., Exeter

Foreword

THE Bangalore Conference on "Science and Technology Education and Future Human Needs" was the result of extensive work over several years by the Committee on the Teaching of Science of the International Council of Scientific Unions. The Committee received considerable support from Unesco and the United Nations University, as well as a number of generous funding agencies.

Educational conferences have often concentrated on particular disciplines. The starting point at this Conference was those topics already identified as the most significant for development, namely Health; Food and Agriculture; Energy; Land, Water and Mineral Resources; Industry and Technology; the Environment; Information Transfer. Teams worked on each of these, examining the implications for education at all levels (primary, secondary, tertiary, adult and community education). The emphasis was on identifying techniques and resource material to give practical help to teachers in all countries in order to raise standards of education in those topics essential for development. As well as the topics listed above, there is also one concerned with the educational aspects of Ethics and Social Responsibility. The outcome of the Conference is this series of books, which can be used for follow-up meetings in each of the regions of the world and which can provide the basis for further development.

JOHN L. LEWIS
Secretary, ICSU-CTS

Contents

List of Figures

1

Introduction

CHARLES TAYLOR

Preliminaries

I am not an information scientist, nor an information technology expert—I am an ordinary university physicist. But in almost 40 years of teaching and research I have come to realise that the transfer of scientific and technological information to children and to the general public is every bit as important as the search for information and its transfer to professional colleagues, which is usually seen as the primary role of the research scientist. Experts in information technology will soon discern that my reading in the field is very limited, but I think that, for the purposes of this volume, which are entirely pragmatic, there is an advantage in having an introduction prepared by an interested observer of the field and a practitioner of some of its aspects, rather than by a theoretical specialist.

There are three sections: first, a rather rambling discussion of what I understand by information and by its transfer and technology; secondly, a discussion of some of the topics that will be covered in later chapters or that were touched on during the workshop sessions of the conference; and thirdly, a rather philosophical section in which I want to suggest some tentative ideas of my own.

This introduction is based on a plenary lecture given at the conference and which was illustrated with demonstrations of several of the techniques discussed; inevitably descriptions of demonstrations do not have the same impact, but they are, nevertheless, better than nothing.

The Information Society

It is said that we live in an information society and that there has been an information explosion during the present century. You have only to look at the size of the current volumes of *Chemical Abstracts* on the library shelves and compare them, or indeed any other volume of abstracts, with those of 75 years ago, to realise the truth of the second statement. But what does the first statement mean—"An information Society"?

Many people concerned with science and technology or with education would probably immediately jump to the conclusion that we mean that we live in an

1

age when the storage, transmission, and retrieval—and indeed the creation—of information by electronic means has suddenly become commonplace, and this is certainly not something that can be ignored.

The dreadful word "Informatics" has been coined; schoolchildren are introduced to computers at a very early age—every primary school in the UK has a microcomputer, though whether this is yet justified in terms of what they do with them may be open to doubt. About 30 years ago a group of learned gentlemen came to the conclusion that two, or possibly three, computers of the size of the Ferranti Mercury would satisfy the computing needs of the scientific and technological communities of the UK for the forseeable future. That must be the understatement of the century!

Clearly our educational system must take all this into account and adjust, both to the possible use of microcomputers in teaching other topics, and to the preparation of students to take their place in the modern world of high technology.

A Historical Note

But what do we really mean by "information" and is the idea of an "information society" so very new? I would like to examine this idea a little more closely.

When, as a tourist, we see a sign saying "Information" or "Renseignement", or "Auskunft", etc., we expect to discover facts. How far is it to the railway station? Where can I park my car? Which is the best theatre in town? What time is the next bus to the beach? But information must surely be more than facts?

Webster's Dictionary gives several definitions, including, for example:

Communication or reception of knowledge or intelligence. Knowledge obtained from investigation, study, or instruction. A numerical quantity that measures the uncertainty in the outcome of an experiment to be performed.

While on the subject of definitions, Webster defines "knowledge" as:

The fact or condition of knowing something with familiarity gained through experience or association.

But whether we are dealing with facts, knowledge or experiences, they can only be meaningful when they are passed on.

So it seems that information must be intimately concerned with communication.

Primitive man had a big advantage over all but his very near relatives among the animals—he had a flexible face and two limbs that were not needed for support, and hence he could make facial gestures and signs. Presumably these were the earliest means of communication, and one could certainly say that a threatening gesture was a form of communication. But the gesture could only be meaningful if both parties were present. The invention of signs and the painting of signs and symbols on, for example, cave walls, carried things a little further; the sign could still pass on its message in the absence of the sender. But it was the invention of portable signs that was the first real breakthrough. I suppose you could say, therefore, that the "information society" was born seven or eight thousand years ago with the invention of the cuneiform system of impressing wedge-shaped marks on a clay tablet and then baking it to preserve it. The preservation was pretty permanent too! As a student I worked on clay tablets of the period 3000 to 2000 BC and found out how the Sumerians measured the area of their fields.

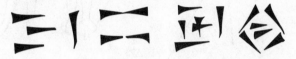

FIG. 1.1. Example of cuneiform script.

Information and Language

So language was born. But verbal languages can also be a barrier if no *common* language is available. From time to time music and mathematics have been described as universal languages.

$$E = m c^2$$
$$s = ut + \tfrac{1}{2} ft^2$$

FIG. 1.2. Examples of so-called universal languages.

But, if you think about it they are not really universal in the strict meaning of the word.

In their book on visual communication in science, David Barlex and Clive Carré quote an interesting instance of an attempt at a universal language. When NASA's deep-space probe Pioneer was launched in the early seventies, a pictorial message was placed on it which it was hoped might be understood by other beings in space.

But Sir Ernest Gombrich, writing in *Scientific American* in 1972, pointed out that this picture can only be interpreted if the observers have a great deal of prior knowledge of certain conventions. The representation of three-dimensional figures in terms of thin lines only works if you have seen such drawings before; the drawing could be taken to be of some pieces of wire bent to certain shapes; the woman's right arm only looks plausible if you know that it is meant to be partly behind her. The arrow only makes sense if you have met a bow and arrow before.

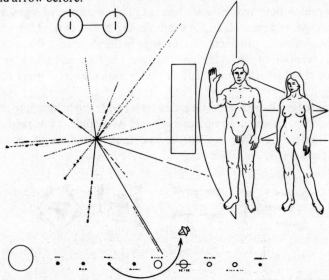

FIG. 1.3. The pictorial message carried by NASA's deep space-probe Pioneer.

Richard Gregory, who has made a special study of optical illusions, has a friend who was blind for many years and then in later life had his sight restored. Richard tells how his friend was unable to "see" many of the common illusions until several months after his sight had been restored. For example, the familiar illusion of equal lines that look to be of different lengths, only works if you are familiar with the corners of rooms or buildings.

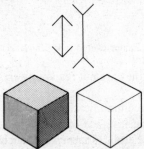

FIG. 1.4. Two optical illusions: *top,* are the vertical lines of equal length? *bottom,* are these hexagons or cubes?

The left-hand picture is like the external corner of a building that is near to you; the right-hand picture is like the internal corner of a room that is away from you. Your brain therefore compensates and makes the vertical line you think is nearer, appear smaller. The cube, even when shaded, looks like a cube, rather than a shaded hexagon, only if you are familiar with the convention.

Languages, perspective, hand signs, all come into the category of agreed conventions; and I suppose the ultimate convention is a digital language, though it is regrettable that computer languages are not all instantly compatible; then the irreducible minimum of prior experience is the ability to turn on the power.

Information and the Human Brain

But some of the information gathering activities in which we indulge involve innate abilities or instincts rather than agreed conventions. Examples would be our brains

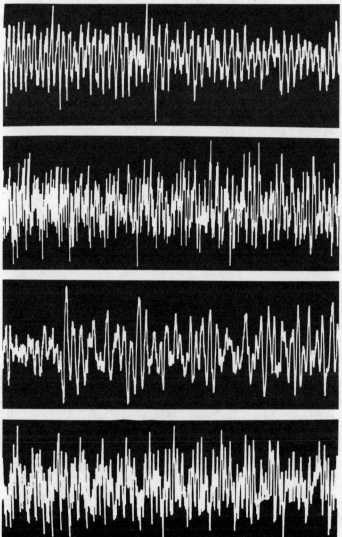

FIG. 1.5. Wave traces of complex sounds: three are of different kinds of music and one is of audience noise.

turning the retinal image upside down, the proprioceptive signals that tell a horse where its back legs are even though it can't see them or enable us to put a key in a lock, the operation of our balance canals that enable us to perform the trick of remaining upright, or the various tricks of our ear–brain system such as to judge the location of a source of sound.

We take many of these ear–brain tricks completely for granted, and yet, without them we should find life very much more difficult. For example, if, during a lecture a baby is heard crying from one side of the lecture room and a dog barks at the other side, while a fire-engine passes outside, we identify the separate sounds. But, unless they are very loud indeed, we are able to ignore them and concentrate on the lecture or, alternatively ignore the lecture and concentrate on one of the other sounds. We use the same sort of mechanism to identify the separate instruments in an orchestra. But this is a very remarkable skill, because all the tiny changes in air pressure that constitute the sound waves from each separate source all add together to become one single wave at the ear or at the microphone; the ear–brain system can perform the analysis though mathematicians would find it difficult.

Again, if we are in an enclosed room, we receive every sound several times because of the reflections from the walls; but, although our ear–brain system could easily distinguish the separate sounds, this would not be helpful and so we fuse them together into a single reverberant sound.

Teaching and the Transfer of Information

We started to talk about the meaning of information, but it seems to me that we have already uncovered a point that is of profound fundamental importance in any sort of teaching or lecturing. It is particularly important when we are thinking of adapting ideas

FIG. 1.6. Cartoon from the *New Yorker*.

or methods to different cultural backgrounds. Our attempts to transfer information to our students *must* be geared to the level both of their prior knowledge of the agreed conventions and of the innate abilities they have acquired through experience. This is a very obvious point, but it's astonishing how often it is ignored. We should not be too surprised if ignoring it leads to some strange consequences.

I remember as a very young primary school pupil attending a singing class and for weeks on end the teacher would scream at us "You're flat! You're flat!" and I used to wonder what she meant; did she want us to be more lively? Did she mean we must stand up straighter? One little demonstration of notes flat, sharp and in tune would have saved her (and us) hours of anguish!

Speaking of demonstrations, obviously a little practical demonstration can make an incredible increase in the speed and efficiency with which information is transferred.

For example, I have been talking about the need for prior experience in the recipients of information and in a lecture on this topic I often use a very dramatic illustration to bring home the point. I have several recordings that can be used; one is of rather poor synthetic speech, one is of greatly speeded-up speech and a third is recorded through a very narrow-band filter so that the sound is distorted quite badly. In each case the audience does not understand the speech when it is heard for the first time; but on hearing it again, after being told what the words are (i.e. after being given the necessary prior experience) they find to their amazement that they can understand it, and cannot imagine how they failed to understand it the first time.

Information and Communication

So, we have begun to explore something of the meaning of information and some of the important conditions to be filled if it is to be efficiently transferred. But there is one further point that I should like to make before moving on to the second section of this introduction. Sir David Attenborough, in his book *Life on Earth*, points out that the genetic code is probably one of the most vital repositories of information in the world. When a species becomes extinct, the code is lost for ever. But he also points out that our libraries, which are the descendants of the mud tablets used as one of the earliest means of storing and transferring information, are really immense communal brains, memorising far more than could be stored in any one human brain. He goes on to say that our libraries

. . . can be seen as extra-corporeal DNA, adjuncts to our genetical inheritance as important and influential in determining the way we behave as the chromosomes in our tissues are in determining the physical shape of our bodies. . . . Cut off from our libraries and all they represent and marooned on a desert island, any one of us would be quickly reduced to the life of a hunter gatherer. . . . Man's passion to communicate and to receive communications seems as central to his success as the fin was to the fish or the feather to the birds.

So the species *Homo sapiens* is a compulsive communicator, and that may be one of the features that distinguishes it from other animals—but we won't pursue that line of thought!

Difficulties in Communication

But one topic that we must pursue and which was drawn to our attention by Marlene Thier in one of our workshop sessions, though it does not appear in a later chapter, is that some people, and children in particular, have disabilities of one kind or another which make communication and learning particularly difficult. She gave us a graphic demonstration of this by asking us to attach a sticky label to our foreheads and then, without using a mirror, or any other aid, to write the word "dabbing" on the label. Try it for yourself: it is a most frustrating experience; you know exactly what you want to write but the complications of working out which way round the letters must be to appear correctly on the forehead defeat most people.

Fig. 1.7. Audience participation.

Children with learning disabilities need our very special attention and we must remember that these disabilities may mask tremendous potential. They often need a different learning style and the skilled teacher must find ways of unlocking the frustration. "Hands-on" science experience is one possible vehicle for the development of language and perception skills. In the same session David Park, of Sinclair Research, demonstrated the "Possum" adaptation of a Spectrum computer that enables physically handicapped people to operate a computer by merely sucking and blowing through a single tube.

This topic provides a convenient bridge to lead us on to discuss the actual topics covered in the workshops rather briefly, because they will be amplified in later chapters.

What Went On in the Workshops

We started by playing the Báfá Báfá game which is a splendid "ice breaker" to speed up the process of getting to know each other and to make contributions to the group discussions. It has a powerful message about the problems and difficulties of being suddenly placed in an alien culture that was particularly relevant to this conference. It, together with another game that was played later on, is described in Chapter 2, which we have entitled "Transfer with Almost No Technology".

Another example of this kind of transfer that was not discussed during workshop sessions but which seems to me to be of supreme importance, especially in International meetings is the break for coffee or tea! I think most people would accept that some of the most valuable interchanges occurred during those periods.

Next we looked at what I described in the pre-conference brochure as "Low-Tech" transfer techniques; this was not intended to be the least bit derogatory—in fact on the contrary, it suggested that, although they are cheap and readily available, they are nevertheless immensely effective. So we renamed Chapter 3 "Transfer with Easily Available Technology—Games".

I'd like to refer to three items. Obviously slides come into this category.

And many people still equate slide shows with the nineteenth-century "magic lantern", or with endless sequences of holiday slides endured by patient friends and neighbours. But if you have not yet seen Mark Boulton's superb tape-slide presentations you have really missed something. And I should like to quote two points from his paper (which appears almost in full in Chapter 3). He warns us against undervaluing the intelligence of the audience and points out that the use of slides does not *automatically make any talk interesting.* He says: "The most effective presentations are often those where the visuals form a 'structured background' to a talk which *would be interesting in its own right— even without the pictures."*

Books, posters and sketches all come into this chapter and I must pick out Yona Friedman's absolutely delightful "matchstick" drawings for special mention. They are simple, carry a powerful message, are easily reproduced or copied locally, and can easily be adapted to suit local languages and circumstances.

At the other end of the scale of sophistication the superb posters produced by the "Man and the Biosphere" project are the ultimate in aesthetic appeal and information content. I should like to quote from the paper prepared by the team, because it seems to me to be of general relevance to our topic of information transfer. They point out that the research results from a project need to be communicated to people at all levels:

Finding the middle ground to allow these diverse audiences to benefit from environmental research means that the results must be "digested", "translated" and "adapted" to their needs. How can scientists ensure that the results of their research are effectively communicated to a lay audience? Who is best qualified to transform the information? Communication specialists, scientists themselves, users or some combination of these? Which are the best media for teaching such diverse groups? Individual scientists and research institutions have been grappling with these questions but, so far, there are no clear answers. Unesco's Man and the Biosphere (MAB) Programme is testing one approach through its poster exhibit "Ecology in Action".

FIG. 1.8. Mark Boulton talking about tape-slide techniques with MAB posters as a background.

Chapter 4 deals with methods of communication between teachers and advisers and was originally merely entitled "Networks". But during the workshop session on this topic Jorge Barojas from Mexico coined a phrase which we liked so much that we adopted it as the title; he said "Networks are just nets that work!"

The whole conference was a direct descendant of the UNCSTD conference, held in Vienna in 1979, and a Unesco Congress held in 1981. The first identified issues that are essential for development and the second stressed the vital role that science and technology education has to play. The Bangalore conference sought to work out practical strategies for actually influencing development in various parts of the world and in Chapter 5, Edward Ploman of the United Nations University has developed the strategy of "Global Learning".

In parallel with the remarkable developments in computer technology that have occurred in the last 10 years or so, there has been a revolution in video techniques. Ten years ago the only way for teachers to produce "movies" for educational purposes was by means of film cameras, with a long wait for the film to be processed and edited; now we have lightweight video cameras which record both sound and pictures for instant playback. Ten years ago the possibility of relaying educational programmes by television throughout a country such as India seemed remote; now we have the satellite programme INSAT in operation. These and other topics are grouped in Chapter 6 under the heading "Transfer using Video Techniques".

In Chapter 7 we move into the sphere of microcomputers and I shall just give three quotations that seem to me to be relevant. Tom Stonier, speaking of the advantages of the computer as a pedagogic tool, says:

Computers have infinite patience. A computer does not care how slowly the user responds or how often a user makes mistakes.

And, following a computer-assisted learning programme in Canada in the late 60s, one girl said that, ". . . the computer was the first maths teacher who had never yelled at her".

The same author says: "The emergence of computer-based education will, over the next two decades, have an impact as profound on Western Society as did the establishment of mass education in the nineteenth century."

But Maurice Edmundson, speaking about the late 80s warns: "In very few cases will the use of the computer, however good the software, be other than a cosmetic addition to the teaching." Both the papers from which these quotations are taken are included in Chapter 10.

Chapter 8 was originally given the working title of "Data Bases", but after some interesting discussions on the precise meaning of this term we decided that it would be more accurately described by the title "Looking After Data using Computer Technology". But, of course, as was mentioned a little earlier, we must not forget the importance of non-computer based information banks such as libraries. One of the major problems in all kinds of information stores is that of labelling it in such a way that it can easily be retrieved. I include one example of an ingenious method of labelling because it is not likely to be met within the standard works on documentation. It is a system of identifying musical themes.

Very often one hears a piece of music and would like to identify it, but most of the standard indexes require either the ability to read a musical score or to transpose a piece into a standard key, all of which skills are beyond many music lovers. In his book *The Directory of Tunes,* Denys Parsons has devised a remarkable scheme which involves no greater skill than being able to write out a tune with an * for the first note, followed by U if the next note goes up, D if it goes down and R if it is repeated. The system can be used by anyone who can hum a tune however inaccurately, and no musical knowledge is required beyond that. Incredible as it may seem it only requires a sequence of up to sixteen such symbols to identify correctly any one of over ten thousand classical themes or over four thousand popular tunes. Perhaps it is not really so surprising when one considers that there are over 40 million possibilities for the sequence; perhaps it really suggests that composers are not as adventurous as one might imagine!

'DUDUU	UUUDD	U	**Mahler** symphony/1 4m 1t(b)
'DUDUU	UUUDD	URDDD	**Beethoven** string quartet/2 in G op18
'DUDUU	UUUDU	UUURD	**Bruckner** symphony/5 in B♭ 2m 2t
'DUDUU	UUUUD	DD	**Paganini** Caprice op1/22 violin
'DUDUU	UUUUD	UDDDU	**Fauré** nocturne/4 in E♭ op36 piano
'DUDUU	UUUUU	DDDDD	**Handel** Messiah: And with His stripe
'DUDUU	UUUUU	DDDU	**Schubert** symphony/1 in D 4m 2t D82
'DUDUU	UUUUU	DDUDU	**Gounod** Mireille: O légère hirondelle
'DURDD	DDDDR	RU	**Max Bruch** violin concerto/1 in Gmi
'DURDD	DDUD		**Mendelssohn** violin concerto in Emi
'DURDD	DDUUU	RU	**Grieg** Lyric suite, nocturne, piano op
'DURDD	DUDRU	UUUDR	**Tchaikovsky** symphony/4 in Fmi op3
'DURDD	DURDD	DDUUD	**Schubert** string quartet/15 in G 1m D
'DURDD	DUUDU	DDU	**Bach** violin sonata/6 in G 2m BWV10
'DURDD	RDUDD	UDUDU	**Mozart** Adagio in Bmi K540 piano
'DURDD	RUUUD		**Verdi** Rigoletto Act I: Figlia!
'DURDD	UDDDU	DDDUD	**Schumann** quartet for piano/strings
'DURDD	UDDUD	DUUDD	**J Strauss Jr** Nacht in Venedig: overt
'DURDD	UDUDU	UDUUD	**Bach** organ sonata in E♭ 2m BWV525
'DURDD	UDURD	DU	**Mozart** Fantasia for mechanical orga
'DURDD	UR		**Ippolitov-Ivanov** Caucasian sketches
'DURDD	UUDDU	UUD	**Debussy** Rapsodie for saxophone/or
'DURDD	UUDUR	RDURR	**Thomas Morley** Fire, fire

FIG. 1.9. Part of a page from Denys Parsons' *The Directory of Tunes.*

One of the exciting developments of the last 20 years or so has been the growth of "distance learning" projects of various kinds, most of which involve a composite package of different kinds of learning materials. We have therefore entitled Chapter 9 "Packages—Learning at a Distance". And in Chapter 10 we discuss the important topic of teaching about information itself. Several schemes are already in operation and we were able to hear first hand about such programmes in Hungary, in England and in Zimbabwe.

Finally, Chapter 10 includes papers that were written for the conference and which we felt should be reproduced as case studies of various aspects of our theme. Also in Chapter 10 is included the results of the questionnaire that was answered by about one-third of the participants.

The Usability of Information

In my last section I want to draw out a few philosophical thoughts that I have tried to formulate before but which were considerably clarified during the stimulating sessions of our group.

The first concerns what I like to call the "usability" of information. Scientists and enlightened members of the general public are familiar with the second law of thermodynamics—a knowledge of which was once described as the hallmark of scientific literacy. It really says that it does not matter how big a store of energy you have, you cannot extract any of it to do useful work unless you have a temperature difference. In other words much of the energy around us is unusable and it is not the total that matters, but the total of *usable* energy. I like to draw a parallel with information. You can have enormous amounts of information stored but what matters is its usability. Why should information be unusable? It could be lost because it has not been efficiently labelled. A book can be lost for a long time if it is replaced on an incorrect shelf in a library, or if the original cataloguer misunderstood the subject; a file can be lost if it is incorrectly labelled or placed in the wrong cabinet; data in a computer information bank can be irretrievably lost if the key words are not skilfully selected. These are obvious examples. But information can also be unusable because it is not in the most useful form. For example, the information contained on an audio tape can be retrieved easily if it is played through a good quality playback system; an inferior system which introduces distortion may make it impossible to understand; and if the information is displayed visually as the wave trace on a cathode-ray oscilloscope I defy anyone to understand it, although the same information is there (See Fig. 1.5).

An illustration that I often use from the field of visual images is to place a slide in a projector but remove the projection lens. The result on the screen is a fuzzy patch of light and it is impossible to deduce from it the nature of the slide. Yet a moment's reflection will show that all the information must be there in the fuzzy patch. When the lens is put back all it can do is to *rearrange* the information, it cannot add anything new about the nature of the slide. The operation of focusing the lens reveals the information, but it is interesting to note that focusing is only making the image look like we think it ought to look. If we are totally ignorant about the object we cannot focus it. Normally of course there are sharp edges to lines, specks of dust, hairs on the slide, all of which we recognise and can use to perfect the focus, but if the object is a blob of biological material under a microscope that nobody has seen before the problem is more difficult.

The Efficiency of Information Transfer

This leads me on to the second general point that I want to make. That is that if the transfer is not made efficiently we may not only miss out some of the information but we may even receive incorrect information. For example, it is

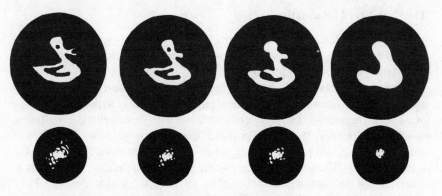

FIG. 1.10. *Top left,* the original object; *bottom row, left,* the diffraction pattern of the object; *remainder of the bottom row,* diffraction pattern with increasingly restricted aperture; *remainder of top row,* images obtained by recombining the limited diffraction pattern from the bottom row.

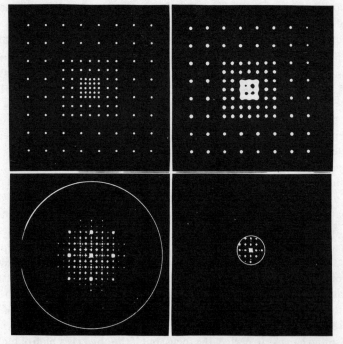

FIG. 1.11. *Top left,* the original object; *bottom left,* its diffraction pattern; *bottom right,* diffraction pattern restricted by an aperture; *top right,* image obtained by recombining the limited diffraction pattern.

possible, using a modern laser diffractometer, to study the diffraction pattern of a complicated object (this corresponds to the fuzzy patch on the screen in the lensless projector experiment) and then to recombine it to form an image of the object (as was done when the projector lens was replaced) but with apertures inserted to restrict the amount of information being fed to the image. This illustrates what happens in an imaging system if the resolution of the system is restricted either accidentally or deliberately. Figures 1.10 and 1.11 show some examples in which quite clearly falsification of the information is occurring. This happens to a greater or lesser extent whenever an image is transmitted through an optical system, or when an audio signal is transmitted through an electronic system.

I suggest that it happens in other circumstances, for example when information is passed on by a teacher or through a textbook or through a video presentation. There is always an aperture and it is our job as educators to see that the aperture is as large as possible and provides as few distortions as possible. Of course, sometimes deliberate restrictions are introduced or unbalanced emphasis is used in information transfer. We call this censorship or propaganda, and it is important to remember that, just as in optical systems, seeing only part of a story may lead to falsification of the information.

Conclusion

We have been talking about a wide range of techniques for information transfer that apply to all levels from pre-school to tertiary education and to communication with the general public. But it is important at the end of this introduction to stress that we have not said that *all* these techniques must be used; nor have we tried to show that anyone is better than any other. In fact I should like to suggest that the key word for this volume is "APPROPRIATE".

If our science and technology education is really to serve future human needs we must ensure that the technique of information transfer chosen is:

Appropriate to the age and experience of the receivers.
Appropriate to the subject matter.
Appropriate to the abilities and skills of the teacher.
Appropriate to the financial circumstances.
Appropriate to the cultural background of the region
and Appropriately free from distortions and confusions.

And that is clearly an appropriate point at which to conclude this introduction.

References

Attenborough, D. (1978) *Life on Earth*, BBC.
Barlex, D. and Carré, C. (1985) *Visual Communications in Science*, CUP.

2

Transfer with Almost No Technology—Games

MIKE ROBSON

THE USE of games in education is fairly well known and very well researched and reported upon (see bibliography). The only problem is that games are seldom used in everyday teaching about science and technology. There may be something of a chicken-and-egg situation here. Because science teachers have not experienced the use of games, they do not use them: because games are not used, we don't gain experience of them.

Accordingly, when we planned this workshop we were determined to break into the cycle and actually play two games. Here are reports on these by players.

BaFa BaFa, a cross-cultural game

We were told that the aim was to create a situation which allows us to explore the idea of culture: to create feelings similar to those one might encounter when travelling to, observing, and interacting with, a different culture.

The overall plan was to divide into two groups in separate rooms. Each group was to learn and practise a different, strange new culture until familiar with it. Then each group would send observers into the other room, to study the other culture and report back. Finally, each group would exchange visitors, who would try to participate fully in the other, alien culture.

This all sounded fairly mad, but we went along with it. As it turned out, however, it was a memorable experience, and I discovered a lot about myself in a short time. I found myself quickly adopting and embracing my new ''culture''. I experienced a traumatic shock on visiting the other ''culture''. At the end of the game I found myself wholly partisan, devoted to my own and rejecting the other ''culture'' as weird and alien. A friend of mine, who had been directed into the other group, expressed feelings which were the mirror-image of my own.

If such firm attachments and such strong feelings can be generated inside a 90-minute game, how much more powerful must be the real-life experience, in which one has perhaps 10–20 years in which to learn one's culture before experiencing another.

The most interesting thing about this game, however, was its effect upon our workshop group as a whole. We had all shared a common experience, at times hilarious but at times somewhat jolting. Perhaps as a result of this shared experience we quickly got to know each other, and started working together at once. There were few defections to other groups, and very little absenteeism in the days that followed. I would recommend this game as an excellent starting point for anyone organising a working-week for a heterogenous group of educators.

The second game, played much later in the week, was one which had been devised and published in Zimbabwe for use in sixth form geography teaching. Here is a report by one of the players.

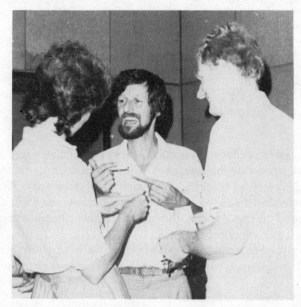

FIG. 2.1. The alpha culture stand close together and talk happily about relatives.

The Mupata Gorge Controversy

I have to admit that at the moment when this game started I thought that Zimbabwe was a village in Thailand. However, by the end I not only knew exactly where it is, but had become passionately aware of the nature of its natural resources and the problems involved in making development decisions.

There were about 40 participants. We were told that the date was 1981. Zimbabwe's economy was expanding at 8% p.a., and planning decisions were urgently needed in respect of new sources of electrical power. Singly, or in pairs, we were assigned important roles, and given slim packs of information about elements of the situation, together with overhead projector transparencies.

After only a few minutes, in which I desperately tried to get to grips with my role of Chief Engineer in the Central African Power Corporation (CAPCO), we were off. One after another we came forward, placed our transparencies on the OHP, and tried to place certain facts before the group.

It soon transpired that Zimbabwe is in fact a small, landlocked African country with an expanding population. At that time (1981) the economy was growing, and it was necessary to double the output of electrical power very quickly. Three possible routes became apparent. We could choose to throw another major dam across the Zambezi river and generate hydro-electric power; three possible sites for the dam were offered. Alternatively, we could build a major thermal power station to provide an equivalent amount of energy by burning some of the vast coal resources currently being thrown away during the extraction of coking coal for the steel mills. Or (for a much smaller sum of money) we could build a power line into neighbouring Mozambique, and purchase electricity from the existing Cabboro Basa project which was producing far more than underdeveloped Mozambique could possibly use.

The techniques of hydro-electric and thermal power generation were quickly explained to us by players. The relative costs were presented, and it all seemed fairly straightforward. The first few players came forward somewhat diffidently to give part of the picture, but then things began to hot up. Someone jumped up and explained that the security situation in Mozambique was very bad, with dissidents repeatedly sabotaging the existing power lines. There seemed little point in investing money in a futile attempt to buy power from our neighbour. It soon became clear that the best option might be a dam across the Zambezi at Mupata Gorge. This would provide the best site, with a vast area of lake for recreation, irrigation, and massive fishing projects.

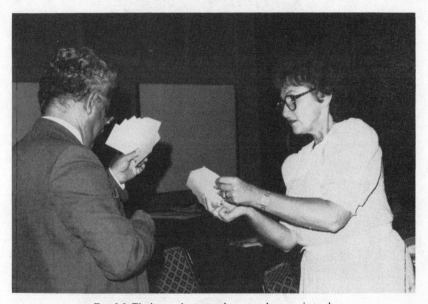

Fig. 2.2. The beta culture stand apart and engage in trade.

Then the controversy began to rage, with upraised voices and shaking of fists. A conservationist revealed that the area proposed for flooding was one of the most beautiful game parks left in Africa. An ecologist explained the delicate balance between very large populations of big animals and the food yielding shade trees alongside the seasonal pools. Once these trees were submerged the animal population would be wiped out. The director of the Zimbabwe Tourist Board, with a very strong Punjab accent, showed the importance of the foreign currency earning potential of the area. Once converted into a lake it would have very little value for tourism, since it would duplicate the well developed facilities of the existing Lake Kariba region, which itself will remain under-utilised well into the next century. The Department of Fisheries official similarly explained that Kariba's potential fish output will not be reached for several decades.

As argument and counter argument were thrown across the tables, the Director of the game kept urging us to make a decision; the human population was growing, he said.

FIGS. 2.3a, b. The beta culture also use a strange language and strange signs.

Without more electrical power for factories and mines and railways, development would slow down and the Government would fall.

At the height of the controversy the Director stopped the game. He explained that at this point the students would then move on to explore the concept and techniques of Environmental Impact Analysis. He invited us to vote on the issue, and by a small majority we elected to build a very big thermal power station instead of a dam. The Director commented that our decision was in fact the same as that taken by the Zimbabwe Government in 1982.

Several things strike me about this game. First, I was amazed at how much I learnt about a very complex subject in a relatively short time. Second, I noted how very enjoyable this learning experience was. Third, I was surprised to discover that this excellent game had been developed and tested by a young lady working at a teachers' college in the middle of Zimbabwe. She did it because she was fed up with using teaching materials imported from North America and featuring North American problems. I think that many of us from developing countries came away from this session determined to follow this self-help example.

In the discussion which followed the game, one lady, a schoolteacher, observed:

I liked it very much. I really believe that this technique is very educative. But in my school we are very examination orientated. Surely, I would not be allowed to spend valuable teaching time in such an activity.

Bibliography

Armstrong, R. H. and Taylor, J. L. (1970) *Instructional Simulation Systems in H.E.* Cambridge: Cambridge University Institute of Education.

Bocock, S. S. and Schild, E. O. (1968) *Simulation Games in Learning.* Beverley Hills: Sage.

Charles, C. L. (1973) *Learning with Games.* Boulder: Social Science Education Consortium.

Cruickshank, D. R. and Mager, G. M. (1976) Toward theory building in the field of instructional games and simulations. *Programmed Learning,* 13 (3), 224–9.

Erickson, K. V. and Erickson, M. T. (1979) Simulation and game exercises in large lecture classes. *Communication Education,* 28, 224–9.

Gann, N. (1976) A teacher's personal view of simulation. *Programmed Learning,* 13 (3), 25–27.

Luginbuhl, I. (1981) *The Mupata Gorge Controversy.* Harare: University of Zimbabwe.

Marsh, C. J. (1979) Teacher education simulations: "the challenge of change" examples. *Brit. J. Teacher Education,* 5 (1), 63–71.

Marsh, C. J. (1979) Simulation Games in Action. Gosford, Australia: Ashton Scholastic.

McAleese, R. (Ed.) (1978) *Perspectives on Academic Gaming and Simulation.* London: Kegan Page.

Megarry, J. (Ed.) (1977) Aspects of Simulation and Gaming. London: Kegan Page.

Musella, D. I. and Joyce, D. H. (1973) Conducting-in-basket Simulation: a Handbook for Workshop Leaders. Toronto, Ontario Institute of Studies in Education.

Owen, G. (1978) Rôle-play in the teaching of comparative education. *Compare,* 8 (2), 175–184.

Shirts, R. G., Dunn, T. P. and Wozniak, P. R. (1976) BaFà BaFà—a cross culture simulation. *Game Reviews,* 7, 471–75.

Tansey, P. (1971) *Educational Aspects of Simulation.* London: McGraw-Hill.

Tansey, P. (1969) *Simulation and Gaming in Education.* London: Methuen.

Zuckerman, R. A. (1979) Simulation helps pre-service students acquire pragmatic teaching skills. *J. Teacher Education,* 30, 14–16.

3

Transfer with Easily Available Technology

CHARLES TAYLOR

I SUPPOSE that the most readily available technology for transferring information from teachers to students is that of the theatre. As a young pupil I remember being "turned off" Shakespeare's plays because the person who taught us was so incredibly dull. Later I was fortunate to have a teacher who really brought the plays to life by acting all the various parts and, suddenly, I realised what marvellous things I had been missing. But what has theatre to do with science and technology teaching? I would suggest that there is a very close connection indeed. For example in several countries (including India), science, technology and drama competitions have been used as a vehicle for stimulating scientific and technological interest.

But there is an even closer connection. The art of presenting a lesson to small children, a lecture to older students or a discourse to the general public is a dramatic one. All lecturers and teachers use dramatic means—sometimes explicitly and sometimes implicitly—to transfer the message.

Lecture demonstration is a special form of dramatic presentation that came to the fore in Britain during the ninteenth century. Faraday was a great exponent of the art, as were Tyndall, Dewar and many others of his successors at the Royal Institution. Very often the apparatus can be extremely simple but, dramatically presented, a very powerful message is conveyed. For example, Professor Eric Rogers uses a slide on which a parabola has been drawn. This is projected on to a screen so that the parabola corresponds to the path followed by a small object thrown upwards at a slight angle, parallel to and just in front of the screen. With practice a coin can be thrown up in such a way that it follows exactly the projected parabola. This very simple demonstration gives a focus to a lesson that will be remembered for a long time, and in remembering the demonstration, the other parts of the lesson are also recalled. It is strange that many well-known scientists have felt that to present science in a dramatic way is somehow beneath their dignity; but times are changing and I sense that the effective use of drama in presenting science and technology is once again moving into fashion.

Another very simple but effective use of cheap materials to convey a powerful message is to use newspaper cuttings. Ilori Alonge of Nigeria described how he cuts out articles from newspapers and uses them as posters or work cards in science lessons. He writes:

FIG. 3.1. The use of newspaper cuttings to make posters or readers.

Nigeria has a fair number of national newspapers and a higher number still of provincial dailies and news magazines. Fluctuating economic fortunes might not permit a definite figure to be quoted as some go out of circulation for varying lengths of time and varying reasons. In recent times, besides carrying news of science that make world headlines, e.g. test tube baby reports, heart transplant, etc., the newspapers have allocated columns, sometimes the centrespread to science and technology news and activities on a weekly basis, e.g. *Daily Times* of Nigeria on Wednesdays. Some have created columns for junior science readers, e.g. *The Guardian* on Mondays. Furthermore, the newspapers have been improving their expertise at in-depth description of useful scientific events, vital scientific cum economic statistics, or at least understand the questions to ask the experts in order to inform the public. Often times, they run feature articles on health, agriculture; oil, gas; chemical; and iron and steel industries. A current major focus is on industrial raw materials derivable from agriculture. It is this contribution of the Nigerian newspapers to information sharing in areas of science and technology that I intend to share with you by displaying some newspaper cuttings in the form of posters. Meanwhile, it contributes a quotient of incidental science and technology education for readers. How many readers take advantage of it, however, is not known.

Content of Newspaper Cuttings

The cuttings have been made out of newspapers published in the last 1 year. The cuttings do not represent a random sample or a geographical spread, rather it represents the broad spectrum scientifically inclined news items which have been reported by Nigeria's most reputable newspapers. In some cases, the news item is of local, national or international significance. At other times, the item is field-specific, e.g. medicine, agriculture, engineering. One other way of describing the items is the form of emotional appeal of its caption, e.g. "scientific breakthrough at local research centre"; hazard, disaster, etc. Thus, the content of the cuttings is diverse but they genuinely represent Nigeria's current levels of awareness, and concern on the topics of agriculture, nutrition, health hazards, superstition and science, raw materials for industry, epidemics, science and technical education afforestation, desert encroachment, etc.

Observation

As of now, the status of the contribution of the newspapers in information on science and technology is not being valued nor co-ordinated, neither has its ultimate potential been given serious attention. Its impact meanwhile hovers on the periphery. It is hoped that in the near future, its potential for science and technology education will be meaningfully explored.

Peter Towse and his colleagues in Zimbabwe have carried this simple idea to a slightly more advanced stage. They collect newspaper cuttings on a particular topic and build them into a booklet of perhaps 16 pages of A4 sheets, which can then be reduced by xeroxing to A5 size for use in class. This is still very cheap and is a means of producing excellent readers. For example, they have produced a series of booklets on chemistry and industry in Zimbabwe, one of which is entitled *Minerals*.

The contents list reproduced below lists the headlines from the cuttings:

CONTENTS

Pupils' introduction.
Value of 1983 mineral production beats all previous records.
Landsat orbits the earth 14 times daily to provide survey info to Zimbabwe.
Zimbabwe mineral sales top $458m.
$430m worth of minerals exported.
Value of mineral production in Z$ million.
Gold losing its lustre.
Huge Chinese boost for Zimbabwean steel exports.
Zimbabwe should have a major steelworks.
ZISCO is base for all manufacturing industries.
Exports will soon earn $100m.
Blast furnaces are heart of ZISCO.
The art of steel making and continuous casting.
600 000 tonnes of coke a year.
Copper in the balance.
Trying to stabilise copper trade.
The changing fortunes.
High hopes for copper.

Drought leads to an 80% cut in aluminium making.
Zimbabwe optimistic on asbestos market.
Asbestos earns $69m, goes to 52 countries.
Zim asbestos industry's safety standards are high.

The introduction for pupils and the list of questions included at the end give a useful idea of how the books are used:

INTRODUCTION FOR PUPILS

A. What Does the Booklet Contain?

Very few chemistry textbooks contain much information on the applications of chemistry, yet more and more people feel that we should be aware of some of the more important of these applications. Where, then, can the information we are looking for be obtained, information that is useful and relevant to Zimbabwe? Quite a lot of this information can be found in articles which appear in newspapers, magazines and journals.

These booklets contain a selection of articles taken from newspapers and magazines available in Harare and other parts of Zimbabwe. In order to make it easier for you to use the material, it has been arranged in six booklets, each booklet concentrating on a single topic or theme. The six topics are:

1. Minerals
2. Oil
3. Natural products
4. Resources
5. Pollution and the environment
6. Financial matters.

B. How Can You Use These Booklets?

There are various ways in which you might use them. If a stock of the booklets is kept in the laboratory or classroom on a library basis, you could use them for general background reading. For example, you might have become interested in the ways in which minerals are used and exploited, in which case you might look through the articles contained in Booklet 1 and choose to read carefully those which seem to be of most interest to you.

You may be asked to take part in a class or group discussion, or in a simulation, concerned with a particular application of chemistry. You will then need to read carefully the appropriate booklet and make notes from the relevant articles. For example, if you were asked to discuss the likelihood of supplying more of Zimbabwe's energy needs from coal, you would look at the Booklets 1 and 4. Or, if discussing some of the environmental issues arising from the country's various mining operations, you would look at the Booklets 1 and 5.

Or again, if you are a member of a group undertaking a project for the Young Scientists' Exhibition, you may find it useful to read through some of the articles as part of your initial preparation for the project. For example, you may be studying the effects of population growth on resources, in which case you would look up articles in Booklet 5.

QUESTIONS

1. How can satellites be used to determine the presence of minerals in the earth's surface?
2. For many years, the price of gold was fixed at US$35 per ounce. The price now fluctuates, having risen as high as US$600 per ounce at one time. What might be the advantage of the fixed price system?
3. Steel is an important alloy, being used to produce a wide range of products. Discuss the importance of steel in the economy of Zimbabwe and one other African country of your choice.
4. Describe briefly how the LD process for the production of steel is carried out in a blast furnace.
5. Although asbestos is an important mineral export for Zimbabwe, there are a number of problems associated with its mining and use. Find out all you can about the possible causes and extent of asbestosis among miners and what can be done to overcome it. Are all forms of asbestos equally dangerous to health?
6. OPEC has made a great effort to control the price of oil. A similar effort to control the price of copper failed. Why do you think this was so?
7. Copper is important to the economies of both Zambia and Zimbabwe. Has there been an equally sufficient emphasis in both countries on agricultural development.
8. How has the recent drought in Zimbabwe affected the production of aluminium? Describe briefly what we can do to overcome these difficulties in the event of another drought.

These two newspaper-based techniques raise the question of difficulties that students may have in understanding scientific and technological literature. This problem has been tackled by Abraham Blum and his colleagues in Israel in relation to the reading of Bio-technical texts. Their paper is reproduced below:

IMPROVING STUDENT'S ABILITY TO READ BIO-TECHNICAL TEXTS

M. AZENCOT and A. BLUM
Hebrew University of Jerusalem, Israel

Introduction

Many science educators want their pupils to be able to read, as adults, articles in professional or hobby publications, in spite of the technical terms and style which are typical for this kind of literature. This goal was emphasised by the Bullock (1975) report. Yet, pupils find it difficult to read these texts, in which many technical terms are rapidly introduced (Hart, 1980). Science books are "hard" to read (Lunzer and Gardner, 1979) and pupils often are unmotivated by the usual genre of scientific reports. Working with culturally deprived children, Markovits and Rosenblum (1973) found that these children preferred a narrative version of an "enrichment reading" chapter in Biology over a descriptive and more "scientific" version. Reviewing the state of the art, Lucas (1983) concluded that informal learning sources can contribute much to scientific literacy.

A strategy to enhance pupils' reading ability and motivation, when reading bio-technical texts, was developed by the Agriculture as Environmental Science (1978) Project in their Let's Grow Plants (1972) unit. According to this approach (Blum, 1982), before attempting the bio-technical text, pupils read a story in which the same biological and agro-technical terms are used as in the technical text, but in the more familiar context of a story with dialogues, and in a style which is apt to create empathy with the characters in the story. The feedback from trial schools showed that the story (''The rose detective'') was very popular with 7th and 8th grade pupils. Teachers reported that this story and similar ones motivated pupils to read the technical texts and that they helped pupils to improve their skills of technical reading. This feedback was encouraging, but lacked rigour to be able to substantiate the claims made by teachers. It also was unclear, if the effect was cognitive (higher reading ability and better understanding) or affective (enhancing pupils' feeling they can master a bio-technical text), or both. Therefore a test was developed and administered.

Test Construction

The test consisted of four subtests.

1. *Subjective-affective subtest*

7th and 8th grade pupils were asked to assess on a 5-point Likert scale—

(a) how attractive they found a given bio-technical test,
(b) how interesting the text was,
(c) how easy or difficult it was to understand the text,
(d) how clear they found the practical instructions in the text,
(e) to what extent they knew what to do, if they had to use the instructions in the text.

In these five questions pupils' rating was subjective and the questions were moving from the affective to the more cognitive.

2. *Understanding bio-technical terms*

Students were asked to explain in simple words ten bio-technical terms, which appeared in the text they had read (e.g. sprinkling, transpiration, cutting, bud).

3. *Understanding bio-technical instructions*

Pupils were presented with a list of ten practical questions related to the preparation of rose plants from cuttings (e.g. ''what is the optimal number of buds? Which cuttings should not be used?''). They were asked to indicate the line in the text, where the answer could be found and to write short answers to the questions. In this way we wanted to check if pupils could find an answer in the text (and not from recall) and if they understood the instructions.

4. *Understanding the sequence in a process*

Four steps in the preparation of rose plants (preparing cuttings, marking the rows, planting and heaping) were put into four different sequences. Pupils had to choose the right sequence, as it appeared in the text.

Experimental Design

A first version of the test was given to two 7th grade classes with 70 pupils, in order to learn if they had any difficulty in understanding the test instructions and in order to eliminate too easy or too difficult questions. These pupils were not part of the main sample. Review of pupils' answers showed that only minor changes were necessary.

The experimental treatment was the use of a story-type reader—*The Rose Dectective* (Blum, 1982). Only pupils in the experimental group received this story and were asked to read it before reading the technical instructions on ''How to grow roses from cuttings''. Pupils in the control group obtained only the technical text.

Prior to this procedure, both the experimental and control groups were tested on their general reading ability, using the standardised reading text developed by Ortar and Ben-Shahar (1972). Both comparison groups were also pre-tested on their previous understanding of bio-technical terms related to the subject matter treated in the bio-technical instructions and in the story. This test was identical with the subtest on understanding bio-technical terms. In this way possible differences in both general reading ability and in the specific subject matter knowledge could be controlled through analysis of covariance.

Since the curriculum unit *Let's Grow Plants* (1972) is used in both 7th and 8th grades (pupils aged 13 and 14), the sample comprised pupils at both age levels. In order to compensate for the different reading levels in 7th and 8th grade, standardised reading scores were used. Fortunately, this was possible, because the reading ability test we used has tables with scores standardised for grade levels and each of three terms. Class level was used as another covariate, in order to investigate its influence on pupils' achievements.

All pupils whose reading level was equivalent to the average of grade five or lower were eliminated from the sample. This was done on the assumption that this reading level was needed to understand the texts and the test questions.

The minimal level for statistical significance was chosen at $p < 0.001$ for the cognitive, objective subtests and at $p < 0.001$ for the subjective subtest which had a strong affective element. The latter was chosen at a lower level, because affective parameters tend to change less than cognitive ones.

The Sample

The final sample consisted of 294 pupils in the experimental setting and 252 in the control group. Pupils in both groups came from the same six schools. Five of these were city schools. Parallel classes from each school were assigned at random to either the experimental or control group.

The percentage of girls and boys was similar in experimental classes (52.3 % girls) and in control classes (55.8 %). Sex was another covariant the influence of which was analysed.

Validation

The questionnaire was content validated by five specialists in reading and rose growing. Construct validity was established by Smallest Space Analysis (SSA; Guttman, 1968) and reliability by alpha coefficients. Technical details can be found in the full research paper (Azencot and Blum, 1984).

Results and Discussion

General Reading Ability was very similar for pupils in experimental and control groups. The mean score for the experimental group was 44.15 (with a Standard Deviation of 6.51) and 44.27 (St. Dev. 5.71) for the control. Mean standardised scores which took into account different class levels, were 72.29 for the experimental group and 71.58 for the control.

Also the pre-tested understanding of terms relating to the specific subject matter taught in the bio-technical text was practically equal: The mean score was 4.42 (St. Dev. 1.56) for the experimental group and 4.39 (St. Dev. 1.44) for the control group. Although this difference is insignificant ($p < 0.001$), standardised scores of general reading ability and specific knowledge of terms were taken in account as covariants in computing the post-treatment data.

TABLE 1

Subtests	Items	Experimental Mean St. Dev.		Control Mean St. Dev.		F-Value
Understanding of						
—Terms	10	6.66	2.23	5.87	2.45	18.28
—Instructions	10	6.56	2.63	5.61	2.80	16.56
—Sequence	1	.79	.40	.65	.47	13.43
Subjective-affective Test	5	3.85	.65	3.70	.59	8.70

Table 1 shows the mean scores for the four subtests: F-values take in account the initial (minor) differences in general reading ability and knowledge of specific subject matter terms.

All the differences between the experimental and control group are statistically significant at the level chosen ahead ($p < 0.001$ for objective, $p < 0.01$ for subjective tests). The differences are relatively larger for the objectively measured cognitive learning categories, and smaller, but still very significant, for pupils' subjective assessment of the ease and attraction of reading a bio-technical text. Reading a story which introduces new bio-technical terms and instructions prior to attempting the technical text proved to have both a cognitive and an affective effect on pupils.

When these data were separately analysed for boys and girls, no statistically significant difference was found between the two sexes in the objective-cognitive subtests. This was true for both experimental and control groups. However, an interesting difference between the answers of boys and girls was detected in the subjective-affective domain. While boys ($N = 111$) and girls ($N = 141$) in the control group reached a very similar score of 3.62 and 3.75 respectively, which is statistically insignificant ($F = 3.21$, in the analysis of covariance), the situation was different in the experimental group. In this group girls ($N = 153$) scored practically the same, 3.74, as their peers in the control group, but boys ($N = 138$) received a significantly higher score—3.96. The F-value for the difference, adjusted for reading ability and previous knowledge was 9.63. This means that only boys were affected by the experimental "story-approach", when asked about their subjective assessment of the ease and attractiveness of the technical text. Such a difference between boys and girls was surprising because no differences between the sexes were found in all subtests of the control groups and in the experimental groups' objective-cognitive tests. A possible explanation is that in the story only boys were participating in the dialogue and that therefore the text appealed only to boys, but not to girls. Yet, both boys and girls in the experimental group increased their objectively measured understanding.

As it could be expected, 8th grade pupils had a higher reading ability and more subject matter knowledge than their 7th grade peers. However, in the affective domain there was no significant difference between pupils in grades 7 and 8—in either experimental or control group.

Conclusions

The findings of this investigation show that story-type chapters with dialogues, presented to lower secondary school pupils before they read a bio-technical text enhance pupils' objective understanding of terms, bio-technical instructions and sequences in a process, as well as their subjective assessment of the ease and attractiveness of the bio-technical text. Therefore, the strategy used deserves further application in curriculum development. At the same time possible differences between boys and girls, and between urban and rural pupils deserve further investigation.

References

Agriculture as Environmental Science (1978) A Curriculum Project, Jerusalem: Ministry of Education and Culture.

Azencot, M. and Blum, A. (1984) Effects of a story-based strategy to enhance pupils' ability and motivation to read bio-technical tests. *Journal of Biological Education* (in press).

Blum, A. (1981) Effects of an environmental science curriculum on students' leisure time activities. *Journal of Research in Science Teaching*, **18**, No. 2, 145–155.

Blum, A. (1982) The rose detective—a strategy to train pupils in reading bio-technical texts. *Journal of Biological Education*, **16**, No. 3, 201–204.

Bullock (1975) *A language for life* (The Bullock Report). Department of Education and Science. London: HMSO.

Cronbach, L. (1971) Test validation. In R. L. Thorndike (Ed.). *Educational Measurement*, 2nd ed. Washington: American Council on Education, pp. 433–507.

Guttman, L. (1959) Introduction to facet design and analysis. *Proceedings 15th International Congress of Psychology, Brussels, 1957.* Amsterdam: North Holland Publ. Co., pp. 130–132.

Guttman, L. (1968) A general non-metric technique for finding the smallest co-ordinate space for a collection of points, *Psychometrika*, **33**, 469–506.

Hart, W. (1980) Reading in Science. *In Language in Science,* Study Series No. 16, Hatfield, Herts: Association for Science Education.

Let's grow plants (1972) Student text (Partial English Translation) prepared by the Agriculture as Environmental Science Project, Jerusalem: Ministry of Education and Culture, Curriculum Centre.

Lucas, A. M. (1983) Scientific literacy and informal learning. *Studies in Science Education*, **10**, 1–36.

Lunzer, E. and Gardner, K. (Eds.) (1979) *The effective use of reading*, London: Heinemann.

Markovits, A. and Rosenblum, Y. (1973) Comparison of two versions of "enrichment" reading material in the biology program for the culturally disadvantaged. In Lewy, A. (Ed.) *Studies in Curriculum Evaluation.* Jerusalem: Ministry of Education and Culture, Curriculum Centre.

Ortar, G. and Ben-Shahar, N. (1972) Reading ability Tests. *Bechinuch Hayesodi 9.* Jerusalem: Ministry of Education and Culture (Hebrew).

Another approach to the problems of scientific literature—this time in physics textbooks—has been developed by Jorge Barojas in Mexico. He finds that the writing of a textbook for themselves by physics students not only provides better books than might otherwise be available but also, in the process, is a very effective means of transfer of information about physics to the students, He writes:

In several Latin-American countries, such as Mexico, physics textbooks for junior and senior High School levels correspond to one of the following categories: they are the results of an important but reduced effort of local scientists: they are deficient adaptations or translations of successful foreign projects, mainly from the United States, or they are improvised collections of conceptual errors written by poorly trained teachers. Very often the programs are surrealistic demonstrations of lack of common sense mixed with a poor understanding of the conceptual development of the discipline. Also, such programs do not match the teaching conditions of the country and are the most evident expression of the conservative traditions of an archaic educational system: old fashioned training, almost zero experience in doing research, deficient background concerning educational matters and very bad infra-structure support.

In this note we comment on the production of didactic materials written by students during their professional education as physicists. This has been an effort to overcome some of the above-mentioned difficulties in physics education, and is an important component of the educational project started almost 10 years ago at the Metropolitan University-Iztapalapa.

The students acquire a better degree of understanding of the subject and creatively undertake a research project on physics education when they are considered as potential collaborators and not merely as passive containers of digested information. Certainly some supervision is required with regard to physics content as well as to editorial and pedagogical matters; also artistic help is necessary in the conception and making of the illustrations and other visual elements. On the other hand, the materials have been published only after trials with preliminary editions and when some training experience in education has given to the students adequate motivation and maturity for writing textbooks.

Rather than proposing a strategy for transforming students into authors, our attitude with respect to teaching has had the following purposes: (1) to motivate the participation of students in the classroom and to stimulate non-formal activities leading to the production of concrete, direct and relatively simple projects on education; (2) to train and to incorporate as assistants some undergraduate students who work in the production of didactic materials (essays, reports, articles on education, chapters for notes, programs for microcomputers, laboratory demonstrations, experiments, slides or transparencies and eventually textbooks or manuals); (3) to test these didactic materials in several contexts and to look for adequate opportunities of improving them: learning activities associated with formal courses, teacher-training workshops, publication of papers for educational journals, booklets and books; (4) to ask for technical and strategical advice and support in the different aspects of the process.

Yona Friedman and his co-workers at the Communication Centre of Scientific Knowledge for Self-reliance, use a delightfully simple drawing technique to transfer scientific and technological knowledge about specific topics to communities in many parts of the world.

They have also proposed the establishment of simple museums of technology and the first of these is located in Madras.

One of the essential features of their work is that the communities themselves choose the topics on which information is needed; For example, how to purify the village water supply, how to improve the efficiency of a cooking stove, etc.

The drawings can be used as posters, they can be copied easily by non-skilled artists, and text in the local language can be added. (It should be emphasised, of course, that the initial story-board type drawings need very careful thought and planning, and a skilled and imaginative artist to execute them.)

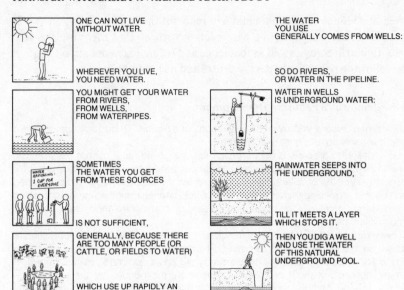

FIG. 3.2. Sample page from Yona Friedman's *Water*.

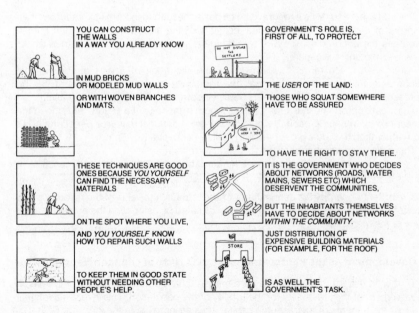

FIG. 3.3. Sample page from Yona Friedman's *The House*.

Often in a village situation a child will read out the captions and adults will begin to learn the words that go with each picture, so that there is a possible contribution to literacy as well as to science and technology education.

The following paper describes the centre and its activity:

1. Self-reliance as a Factor of Development

Development means improving life conditions of a nation. To do such improvement involves inevitably government expenditure.

In countries with precarious economic conditions government cannot assure basic commodities (food, health care, housing and various services) in sufficient quantities for the largest part of the population. In such countries the solution might be approached through people producing themselves their own subsistence in kitchen-gardens, making themselves their utensils, their house and keeping up themselves their health care and community services.

A policy facilitating self-help subsistence and self-help public welfare represents a significant capital saving for a government. The capital thus saved can (and has to) be employed for upgrading the national economy: as basic commodities for the whole nation are obtained without cash investment and are produced by the subsistence efforts of the population, the disposable cash capital can be invested into the development of more sophisticated means of production.

2. Economic Capacity as a Function of Available Resources

The state of a country's economy can be better described in terms of resources than in terms of cash. The national capital consists of the sum of all natural and human resources the nation disposes of (raw materials, land, energy sources, human labour force and human intelligence).

Referring to such terms, investment means allocation of a part of the national resource-capital for a particular goal; the rentability of such an investment can be represented by the rate of the benefits obtained compared to the investment itself.

The most appropriate indicator for the economic capacity of a nation, referring to the concept of the resource-capital, would be the gross national resource per capita (GNR p.c.) rather than the usual gross national product per capita (GNP p.c.).

This indicator corresponds to the following formula:

$$\text{GNR p.c.} = \frac{\text{natural resources} + \text{human labour force} + \text{human intelligence}}{\text{number of citizens}}$$

N.B. In order to quantify the denominator of the formula, the value of the resources, the labour force and the intelligence can be approximated by estimating the present market value of the mentioned items.

3. Development as the Improvement of the Quality of Life for a Nation

The progress of development of a nation can be interpreted as the overall improvement of the life-quality for the totality of the population. This involves a rational share of the benefits of the resource capital, thus an increase of the indicator GNR p.c. For a nation with expanding population the GNR p.c. does not increase but in the best case it can remain stable (indeed, the numerator, which includes the labour force, grows to the same extent as the denominator, the size of the population). But, in order to increase the value of the GNR p.c. in the case of a growing population, the numerator should be increased substantially. Now, this cannot be achieved otherwise than by increasing the human intelligence factor.

Thus the only way to make progress economic development goes through the efforts to increase the national intelligence. Such increase of the national intelligence means, obviously, the increase of the average level of knowledge rather than the increasing sophisticated knowledge of a small part of the population.

The knowledge enabling people to come up themselves with what they need for life can be the best tool for achieving real national development with investing the minimum of cash.

4. The Role of the Communication Centre of Scientific Knowledge for Self-Reliance

The transfer of that kind of knowledge enabling the most disfavoured groups to assure their own subsistence and to improve their life, that transfer of knowledge represents the task the Communication Centre of Scientific Knowledge for Self-Reliance is mandated with.

The Communication Centre of Scientific Knowledge for Self-Reliance is an Institute in the framework of the United Nations University. The mission of the Centre is to make accessible for the disfavoured such knowledge of scientific origin which might enable them to improve their life conditions, making use for that purpose of those means they can dispose of without cash expenses.

In order to transfer to largely unschooled people that kind of information the Communication Centre of Scientific Knowledge developed successfully several particular forms of presentation. Without the work invested by the Centre many people would not easily arrive at that knowledge which, as we saw, is the only implementable path to development.

The particularity of the operation of the Communication Centre of Scientific Knowledge for Self-Reliance is thus to assure the missing link between important scientific knowledge and innovations of any origin, and the poorest, often illiterate people who could make use of the results of those achievements and innovations in a fruitful way.

The staff animating the Communication Centre of Scientific Knowledge must combine competence about new scientific achievements with substantial sociological experience in order to guess right what innovations might be accepted by the target public. Besides that, the same team has to be expert in communication techniques and on media, sophisticated or simple.

In order to be able to disseminate useful information we have to know where to find it. Once the appropriate scientific information is found, we have to adapt it to the social context, to transform it into a form of presentation accessible and familiar to the particular target, the public, and we have to do this in the least expensive manner.

The communication techniques implemented by the Centre include inexpensive presentations like wall-journals or posters, more sophisticated ones like animated cartoon features, and, finally, some highly complex operations like the Museum of Simple Technology.

5. The "Manuals", Base of the Operations

The Communication Centre of Scientific Knowledge for Self-Reliance operates by transposing available knowledge matter into "manuals", largely visualised explanations which form the base of all of our forms of presentation.

The manuals are highly simplified communication supports, with many drawings simple enough to be easily copied by unskilled people, and with little and very simple text as captions. The presentation has to appeal to knowledge already possessed by the target public and has to emphasise the effective benefit they can expect from the innovation exposed.

The least expensive use of a manual implies it's being presented in the form of a "wall-journal". Global costs for such a presentation are only a few dollars for a village and for one topic.

In order to be able to make a manual, the most difficult work is to conceive the "story board" which has to take into account the main features of the technique or method to be exposed and the skills and resources our target public disposes of and is ready to deploy. The techniques we expose appeal more for investment in labour than in cash or in materials which have to be bought. In India, for example, where most of our operations take place, even scrap metal can be too expensive for the disfavoured.

6. Manuals in Audio-visual Presentation (animated cartoons)

The manuals conceived by the team of the Communication Centre of Scientific Knowledge for Self-Reliance are adapted to be presented as animated motion pictures, to be used for diffusion by local television or to be projected at the occasion of training courses on the site. The shooting technique implemented for making these pictures can be employed even by unskilled people and with relatively simple equipment. That particular technique, for which the Centre possesses the exclusive rights, obtained in 1962 the Great Award Golden Lion of St Marc, the highest distinction of the famous Venice Film Festival in Italy.

Animated motion pictures based on the manuals conceived by the Communication Centre of Scientific Knowledge can be particularly valuable support material for field workers of development agencies.

7. A Museum of Simple Technology

A project of a Museum of Simple Technology was conceived, in order to make possible the demonstration in the form of tangible objects the material results of the techniques and methods which serve for topics of the manuals made by the Centre. As those techniques and methods were designed for the use of the disfavoured in order to enable them to improve their life conditions, it is important to present for them some prototypes of those artefacts the manuals show in drawings.

The topics presented in the Museum are the following:

On food growing and preparation:
- gardening on shelves
- minimal kitchen gardens
- food storage and conservation
- food enrichment by fermentation and germination.

On water collection, water storage and conditionment:
- simple water filters
- drip irrigation
- prevention of leakage and evaporation.

On accessible energy sources:
- firewood crop for cooking
- fuel-efficient cookers
- windmills for low-velocity wind.

On health care:
- safe drinking water
- fighting dehydration
- preventing epidemics.

On environment:
- impacts of environment on people
- impact of people on environment.

On enterprise:
- how to start a tea shop (for example)
- how to ask for a bank loan.

The museum-building itself—a cluster of hut-size units—will demonstrate those building techniques which can be implemented by the disfavoured to improve their dwellings.

We are looking ahead to promote a network of such Musea of Simple Technology in many places in developing countries, to be materialised with the co-operation of government agencies, non-governmental organisations or community groups. The Madras Museum of Simple Technology, work on which is due to start this summer, should serve as prototype for the network.

The exhibits should form a permanent feature continuously updated. The Communication Centre of Scientific Knowledge for Self-Reliance, with its parent organisations, UNU and ICSU, will provide the Museum with appropriately presented new scientific ideas.

The Museum of Simple Technology has the role to emphasise the concreteness of the topics explained in the manuals, topics which relate to methods and techniques based on scientific progress and at the same time inexpensive enough to be implementable by the disfavoured.

8. The "Minimal Kitchen Garden" Project

As increased self-reliance facilitates national development, to increase the ability of disfavoured people to sustain themselves and to produce at least a part of their livelihood might be the most effective aid to development. In order to improve that ability, the Communication Centre of Scientific Knowledge for Self-Reliance programmed the Minimal Kitchen Garden Project.

The basic idea of the project comes from the consideration that $2-6$ m^2 for a person might be an area sufficient to create minimal kitchen gardens and thus for growing food plants providing those substances (vitamins, proteins, etc.) the absence of which is dangerous to health and which are practically lacking from poorest people's diet.

On the other hand, $2-6$ m^2 for one person represents an area which can be found not only in rural settlements but even in urban slums, particularly if the right techniques and the right plants are used. For example, gardening on shelves 2 metres long corresponds to about 3 m^2 of effective garden area.

The Communication centre programmed to convey all available information necessary for the materialisation of such minimal kitchen gardens by the rural and urban poor. That implies competence in various subject matters like botanics, dietetics, appropriate technology and physical planning, all of which has to be formulated accessible to the quasi-illiterate.

9. Present Diffusion

The operation of the Centre is very economic. 83.5 % of the funds are invested into effective production, with all other costs kept exceptionally low.

The Centre's material was particularly well received in India. Numerous topical manuals were diffused at All-India level, many of them were taken over by local NGOs, translated into local idioms. Feedback is very encouraging, and many offers from voluntary translators and from field workers come in.

The estimation about the Indian diffusion shows that our manuals circulate in some 100,000 copies. We can consider, conform to local custom about 10 readers at least for one copy, which means that our manuals reach about 1,000,000 people in India. This is a pretty low figure compared to the size of the country, but it is a figure which cannot be neglected. The next phase of our programme will be diffused by the Adult Education Programme of the Government of India, that would mean an audience of at least twenty-fold.

At the suggestion of some high personalities in India we will prepare in the coming year animated motion pictures based on our manuals.

10. Post Scriptum

People not profoundly acquainted with some of the real problems which make poverty a lasting curse of our world, the operation of our Centre seem marginal. Only those who know that development of a better life for the disfavoured goes through access to useful knowledge appreciate our work up to its real value. Among others,

LAO TSE TOLD:

IT IS NOT ENOUGH
TO GIVE A FISH
FOR THE HUNGRY.

IT IS BETTER
TO TEACH HIM
HOW TO CATCH
A FISH.

BUT
WHAT TO DO
WHEN MANY MILLIONS
ARE HUNGRY?

IT IS BETTER TO
TEACH THEM HOW TO
PRODUCE THEIR FOOD
THEMSELVES.

THIS IS OUR TASK.

FIG. 3.4. An example of a proposed illustration for the tutor manuals in the
Adult Education Programme of the Government of India.

Sir John Kendrew, Nobel Prize Laureate and President of the International Council of Scientific Unions wrote about the Centre's manuals: "I have just seen the preliminary materials prepared by the Communication Centre of Scientific Knowledge for Self-Reliance and they seem to me to be among the most exciting materials in the Information Transfer field that I have seen for many years." And this is only one among many other opinions of personalities who think that the Centre's work is one of the "most positive efforts fighting misery".

The use of pictures of a slightly more sophisticated kind to transfer information to village communities has been used in China. Economy in the use of firewood in cooking stoves is obviously important, but the technical drawings may not be readily interpreted by a local builder. But a coloured drawing which can literally be taken apart to show the construction makes the whole design clear.

An even more sophisticated approach to the transfer of information pictorially has been developed by Unesco's "Man and the Biosphere" team. In this case the problem is to transfer the results of specialised research to people outside the specialist area.

The following paper describes the programme.

ECOLOGICAL INFORMATION FOR NON-SPECIALISTS:

AN INTERNATIONAL CASE STUDY
J. DAMLAMIAN and M. HADLEY

Communicating Research Results: for Whom and How?

In the last decade or so, there has been a very marked heightening of awareness to environmental concerns. Coverage by the mass media, together with public and privately sponsored information campaigns at national and international levels, have succeeded in giving the public a sense of the extent, seriousness and urgency of the environmental problems in the world today. This is no small achievement, given the nearly total ignorance, 15 or 20 years ago, of such matters which most people take almost for granted today. And given the difficulty of "competing" with the masses of information being bombarded at those with access to modern mass media.

How have these basic messages about the environment been transmitted? Largely by appealing to people on an emotional level, using striking pictures or graphics intended to make people stop, look, and think. But such an approach, however effective, has limited educational impact, as little or no substantive information is conveyed.

Yet a large body of scientific knowledge about the environment already exists. And research is providing new information all the time. But most of this information is accessible to scientists alone, as it is usually couched in technical language and presented in a way that the layman can neither appreciate nor understand. The need to educate non-specialists about the environment by finding a middle ground of communication—a middle ground between emotions and technical information—is a major challenge facing scientists and communicators today (Fig. 3.5). The goal is clear, but what we do not know is how: How to convey more than just a plea for caring about our environment by transmitting to the layman some of the substance of what scientists already know about how natural systems function.

Ecological research is intended, like nearly all research, to extend the frontiers of knowledge. But the results of such research also have a practical application outside the academic world. The potential users of ecological information .thus transcend the scientific community.

One major user group are those responsible for deciding how natural resources will be used—that very heterogeneous group known as "decision-makers". Decision-makers include the legislators, administrators, industrialists, farmers, fishermen and foresters who decide how much forest will be logged, where a dam will be built, what area of land will be cleared, how many animals will be run, how much money will be allocated to particular areas of research. Also interested in the results of such research is a variety of groups concerned in one way or another with man's interaction with the environment, both locally and in other parts of the world. Educators (ranging from primary schoolteachers to professors and curriculum planners), environmental groups, local associations and the general public are among such potential users of research results.

Finding the middle ground to allow these diverse audiences to benefit from environmental research means that the results must be "digested", "translated" and "adapted" to their needs. How can scientists ensure that the results of their research are effectively communicated to lay audiences? Who is best qualified to transform technical information? Communication specialists, scientists themselves, users, or some combination of these? Which are the best media for reaching such diverse groups? Individual scientists and research institutions have been grappling with these questions, but so far there are no clear answers. Unesco's Man and the Biosphere (MAB) Programme is testing one approach to dealing with these questions through its poster exhibit "Ecology in Action".

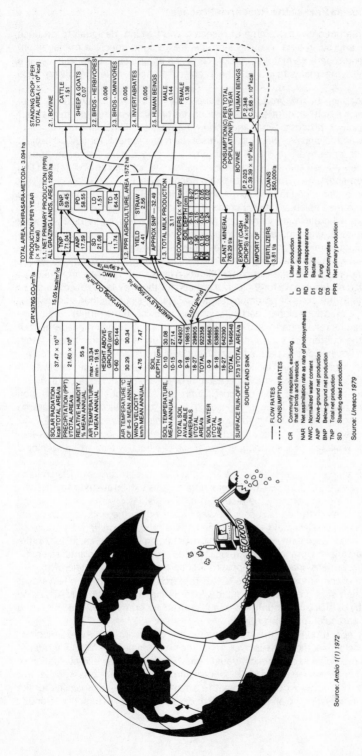

Fig. 3.5. Two ends of an information spectrum. At left, graphic for alerting public opinion to an environmental issue—immediate impact, designed for a general audience, low information content. At right, figure for communicating complex scientific data to other specialists—considerable time required for even a specialist to assimilate data presented, information incomprehensible and indigestible to non-specialist. The graphic and the figure are well adapted to their desired aims. But something else is required to convey to the layman some of the substance of what scientists already know about how natural systems function and react to manipulation by man. The Ecology in Action exhibit is one attempt.

Communication as Part of the Research Process

MAB is an international programme of research and training, designed to produce information for improving land use. Research centres around concrete management problems bringing together whatever disciplines of the natural and social sciences are needed to cover the many facets of the problem at hand, often involving local populations.

From the inception of the programme in the early 1970s, the communication of technical research results to different user groups has been considered an integral part of the MAB research process. Since the underlying goal of MAB research lies in its ultimate application to land use planning and resource management, for a MAB project to fulfil all of its objectives, its results—as well as being diffused through normal scientific channels—need also to be presented in a form which can be understood and used by the "non-specialists" for whom they are also intended.

As MAB field projects began to yield results in the late 1970s, the emphasis on this aspect of programme implementation increased. The initial impetus came notably from the national level, where in a number of countries MAB scientists, by virtue of the very nature of their work, were increasingly interacting with local and national decision-makers, curriculum planners, etc. In the process they encountered many difficulties and challenges involved in communicating the results of ecological research: scientists generally not encouraged to consider communication as part of their research function and not rewarded professionally for their efforts; need for the knowledge of scientists to be articulated with the skills of communication specialists, since neither scientist nor communicator can do it alone; particular importance of transmitting ecological information given its relevance to issues ranging from the day-to-day management of resources, to the content of school curricula and the future of the biosphere itself; communication not included as part of research budgets and make-up of project teams.

The MAB Secretariat within Unesco also began to take a number of initiatives to diversify the kinds of information materials being produced on MAB research. In 1978, the MAB Audio-visual Series was launched, with the first slide–tape presentation dealing with man and the humid tropics (Unesco, 1978[1]). At about the same time, a large colour wall-chart on MAB research in the humid tropics was produced.

It was intended mainly to show scientists how MAB was working to develop interdisciplinary research projects and to link these projects within an international network.

In June 1979 a large-scale exhibit on the Mexican biosphere reserves of Mapimí and La Michilia was held at Unesco Headquarters in Paris. This exhibit tested the idea of a participating country "exporting" the results of a MAB field project to its sponsoring organisation for demonstration purposes. It was organised by the scientists and local populations working and living at these sites, and it was presented in Paris by decision-makers closely associated with establishing and maintaining the reserves, including the Governor of the State of Durango. The exhibit included live animals and plants, scientific charts and maps, professional photographs, rock specimens and local handicrafts.

An exhibit giving an overall view of MAB research was organised in November 1979 for the sixth session of the International Co-ordinating Council of MAB, using materials sent by participating countries. An audio-visual programme included films and slide shows on particular research projects.

These first steps, modest as they were, were the precursors of a more ambitious undertaking to help mark the tenth anniversary of the programme. The MAB Secretariat was asked to prepare an exhibit which would contribute to a review and evaluation of what had been accomplished within MAB during its first 10 years. The idea was that this exhibit would be presented in conjunction with a scientific conference organised for the same purpose. Initially it was planned that the exhibit would consist of national "stands" or displays to give each participating country the opportunity to prepare and display materials on its own MAB research projects. This approach was intended to stimulate MAB scientists, if they had not already done so, to address the issue of communicating their research results to lay audiences. Very heterogeneous components would have been centralised and woven together into a single exhibit to be held in a given place at a given time.

But it soon became clear that such an exhibit would not only be difficult to organise but would also have limited impact in terms of the numbers of people who would benefit from it. In the end, it was decided to try a different approach.

Multiplying the Impact

Rather than prepare an exhibit which would be shown, as most exhibits are, in one place and at one time, the MAB Secretariat in collaboration with a Paris-based group of communication specialists,[2] decided to prepare an exhibit which could be reproduced in multiple copies. Multiplying an exhibit usually implies making two or three copies which "travel" within a relatively limited geographical area. But this traditional approach did not suit the needs of an international programme with over one hundred participating countries, having diverse linguistic and cultural characteristics.

It was therfore decided to prepare an exhibit which would attempt to reach a very large number of people throughout the world, and which would stimulate further efforts on the part of MAB scientists to communicate research results by providing a high-quality core around which larger exhibits could be built. This was the genesis of the Ecology in Action poster exhibit (di Castri et al., 1982a, 1982b; Lefevre, 1984; Unesco, 1981).

The Ecology in Action exhibit (Figure 3.6, Table 2) consists of thirty-six colour posters, printed initially in 1000 copies (and later reprinted in 2000 copies) and in three languages (English, French and Spanish). Such a "poster exhibit" has the potential of reaching a greater number of people than almost any medium short of television, radio and other means of mass communication. A book, for example, if mass produced, may be read or looked at by 40,000 or 50,000 people. A single copy of an exhibit on display for several weeks will be seen by several thousand visitors. Once mounted, an exhibit is "permanently" ready for consultation by the passer-by, unlike a book that has to be opened or a film that has to be shown.

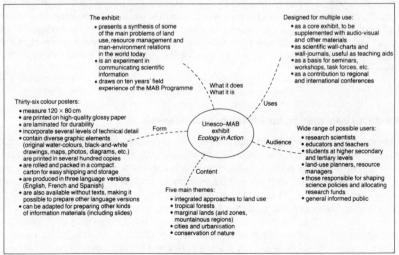

FIG. 3.6. The ecology in Action exhibit at a glance. From di Castri et al. 1982a.

Table 2. Ten design features of the Unesco poster exibit "Ecology in Action".

— Root educational and information materials on the results of field research.

— Seek middle ground between graphics for heightening environmental awareness and ways in which scientists usually report research findings.

— Look for ways to multiply the impact of synthesised information (e.g. multiple copies, multilevel audiences, adaptability to different uses).

— Enjoin the efforts of scientists and communication specialists.

— Present scientific information at different levels of technical detail.

— Shrink the time delay between generation and application of research information.

— Seek balance between scientifically proven information and doubts and uncertainties of science.

— Combine information from social and biological sciences.

— Take account of diversity of geographical scales, time frames and perceptions in respect to land use problems.

— Prepare materials which provide a global perspective within which local problems can be placed.

Multiplying an exhibit by a factor of several thousand further increases the number of people involved. When any of the 3000 copies are used several times, the impact is even greater. The availability of the MAB exhibit in several languages, and the possibility of easily producing other language versions, are further advantages which increase the usefulness of the exhibit on the international scale. From a cost-benefit perspective, the cost per user is very low, especially as the exhibit has been prepared by an international organisation on a non-commercial basis.

This multiple impact approach concentrates scarce funds on communicating the message rather than on the cumbersome and expensive "hardware" (display equipment, etc.) used in traditional style exhibits, hardware which does not always contribute significantly to getting the message across. The poster form minimises such costs. Posters can be displayed in a variety of inexpensive and simple ways (from being stuck on the wall to being mounted on cardboard or wooden panels), depending on local resources. Obviously, this low-cost approach does not preclude a more elaborate presentation limited only by the imagination of the organisers and the money at their disposal.

Multilevel Information for a Variety of Audiences

Identification of a target audience is a key step in the communication of the results of scientific research. In the planning of Ecology in Action, the idea of concentrating on a single target audience was considered at length but finally eschewed. Rather, it was decided to experiment with an exhibit that could appeal to several audiences and thereby further increase its impact. Such an exhibit could later be adapted to the needs and interests of more restricted audiences by adding complementary components. The target audiences of Ecology in Action thus reflect the heterogeneity of the potential users of the results of environmental research: the informed general public, decision-makers responsible for determining science policy and the allocation of research funds, land use planners and resource managers of various kinds, educators and students at secondary school and tertiary levels, research workers from various disciplines, etc.

To reach these different groups, a "multilevel" or "menu à la carte" approach was adopted, multilevel in terms of the user's interest, scientific knowledge and the time available for viewing the exhibit. These factors are vital to determining receptiveness and the user's ability to understand and absorb the information presented. In practical terms, this approach meant incorporating into the posters different hierarchies of information, more or less detailed, more or less technical, more or less time-consuming to read. Individual viewers can thus make use of the exhibit at a level appropriate to their interests, training and available time.

Resolving the dilemma of how much information to put in each poster was not always an easy one. A balance had to be struck between presenting small amounts of easily understood information (possibly undermining the scientific impact and use of the posters as teaching aids) and including many more words and details (and thereby running the risk of losing the interest of the decision-makers and the general public). Some aspects of this dilemma are outlined in Table 3.

Table 3. The dilemma of "How much information?": effects of more or less text (From di Castri *et al.* 1982a)

Less text	More text
Little information content	Information content relatively high
Tendency to resemble advertising material	Tendency to resemble scientific wallchart or poster lectures
Rapid and direct impact	More effort, more time required to digest information
Short "half-life" (e.g a few months–1 year)	Longer "half-lfe" (1–5 years)
Danger of being considered by scientists as too simplistic	Greater chance of capturing interest of scientists
Particularly useful for decision-makers, general public	Less appropriate for communicating with decision-makers, general public
Not particularly useful for teaching	Useful for environmental education and other teaching purposes

The time factor was also a difficult hurdle. The way the information has been presented makes it possible for users to obtain something from each poster even in a very short time (30 seconds), while still including the kind and amount of information that can be absorbed only in a longer period (e.g. 30 minutes). The latter is necessary if the posters are to be used as teaching aids.

The multilevel approach was also reinforced by the need to capture the interest of the various audiences that are likely to be attracted to different styles and forms of visual expressions. Presentation of the results and experience of MAB research for a range of potential users thus called for the use of very diverse visual elements, including: simplification and redrawing of graphics originally presented in highly technical form, use of hard data acquired in field studies to demonstrate general ecological principles, development of new visual representations of certain fundamental concepts underlying MAB research (e.g. human use system, interdisciplinarity, new approaches to conservation, cities viewed as ecological systems), use of photographs contributed by MAB scientists and of maps developed as tools for advancing scientific work. It can thus be said that the Ecology in Action exhibit is indeed "rooted" in MAB research and as such represents an experimental endeavour to communicate technical research outputs to non-specialists (as opposed to the—obviously no less valid—approach of developing educational materials on various environmental subjects without basing these materials directly on information acquired in field research).

Putting the Exhibit to Use

The English, French and Spanish versions of the exhibit were initially printed in a total of 1000 copies. This soon proved to be too limited a number to respond to the widespread interest which had been generated worldwide. In view of the demand, a reprinting was carried out in 1984. As of early 1985, over 1500 copies of the exhibit had been distributed to 122 countries; an indication of how the exhibit has been put to use in a sampling of these countries is given in Table 4.

The many different ways in which it has already been used would seem to indicate that Ecology in Action is a flexible and adaptable communication tool which can be an effective trigger for further action. And the diversity of those expressing interest in the exhibit can be taken as a sign not only that there is a clear need for such materials, but also that the posters do appeal to a variety of audiences. Among those that have acquired the exhibit are secondary schools, universities, research institutes, environmental associations, conservation groups, senior citizens' organisations, prisons, municipalities, Unesco Clubs, National Commissions for Unesco, MAB National Committees, learned societies, departments of forestry and land use planning, ministries of education and, of science policy planning, museums and international scientific organisations.

In June 1982, posters from the MAB exhibit were widely used in events to commemorate World Environment Day and the tenth anniversary of the Stockholm Conference on the Human Environment, in such countries as Argentina, Australia, Belgium, Chile, Cuba, Ecuador, Malaysia, Malta, Mauritius, Mexico, Nepal, New Zealand, Peru, Senegal, Spain, Singapore. In several countries, including Cameroon, India, Netherlands and Portugal, single copies of the exhibit have been mounted and have been sent on tour from one town to another.

Table 4. Sampling of ways in which the Ecology in Action exhibit has been used.

Brazil/Portugal: plans to jointly translate posters into Portuguese and to reproduce in multiple copies.

Chile: used to inaugurate refurbished Museum of Natural History in Santiago; preparation of new panels based on MAB field projects—in mountain areas, grazing lands, continental waters.

China: translation of 36 posters into Chinese; preparation of ten new panels on Chinese MAB research; travelling exhibit to seven major cities; plans to put on permanent display at the planned Hangzhou Centre of Man and Environment.

Finland: preparation of slide package based on posters and distribution by MAB National Committee.

France: major 2-month exhibit for general public at Paris City Hall; preparation and printing of twenty-five new panels on examples and subjects of particular interest to the French public, including some panels based on MAB field projects (both at home and abroad) involving French scientists.

German Democratic Republic: printing of German version of posters envisaged in first quarter of 1985.

India: mounting of several sets of the exhibit and display with locally produced wall-charts in 46 workshop-exhibitions in 14 cities, were viewed by many thousands of schoolchildren. Other uses include display in National Museum of Natural History, during Parliamentary Environmental Forum, on World Environment Day.

Indonesia: translation of five posters into Bahasa Indonesia; printing in several thousand copies and distribution to schools envisaged.

Ireland: used as a roving exhibition in conjunction with a national poster competition for schoolchildren run by national association of art teachers.

Italy: translation into Italian under way, with printing in one thousand copies foreseen; translation will consist also of "experimental" adaptation of existing texts for schools and preparation of teachers' guide.

Ivory Coast: travelling exhibit to major urban centres: displayed on occasions of meetings of Ministers of Scientific Research (1983) and of MAB Committees (1984) of francophone African countries.

Mexico: displayed during Unesco Week in Mexico; used in conferences and seminars organised by Institut di Ecologia, INIREB and other tertiary institutions.

Morocco: used in round tables and travelling exhibitions on themes such as Soils and development, organised in co-operation with the Centre Culturel Francais.

New Zealand: used in training programmes for national parks staff and during the fifteenth Pacific Science Congress.

Nigeria: incorporated in National Science and Technology Fair; used as teaching tool by International Institute of Tropical Agriculture.

Norway: translation of fourteen posters into Norwegian, printing in several hundred copies, and distribution by UN Association of Norway.

Senegal: on permanent display at the Institut des Sciences de l'Environnement; displayed in regional capitals to mark World Environment Day.

Spain: translation of six posters into Catalan, printing in several thousand copies and distribution to all secondary schools in the region; translation of twelve posters into Basque, printing in more than a thousand copies, and distribution to schools in the region.

Sri Lanka: preparation of hand-calligraphed sets of the exhibit in Sinhala and Tamil, loaned to schools by the national authority responsible for Natural Resources, Energy and Science.

Syrian Arab Republic: Arabic translation of text of posters prepared by the Arab Center for Study of Arid and Dry Lands (ACSAD), located in Damascus (possible ways for publication of Arabic version currently being examined with ALECSO).

Tunisia: used in events to commemorate la Journée de l'Arbre ("Tree Day"); on display in the ecological museum of Ickheul (biosphere reserve and world heritage site).

USSR: translation of 36 posters into Russian; preparation of fifteen new posters on environmental protection, printed in 60,000 copies and distributed to schools with a teacher's manual; preparation of special exhibits for First Biosphere Reserve Congress (Minsk, September 1983) by the Soviet MAB Committee (concerning the seven biosphere reserves in the USSR) and the Byelorussian MAB Committee (on the Berezinsky biosphere reserve).

United Kingdom: used in displays at Royal Geographical Society in London, Royal Botanic Gardens at Kew, UK United Nations Association Centre, etc.; placed in the libraries of museums such as the British Natural History Museum and Leeds City Museum.

USA: used in conjunction with World Fair in Knoxville, Tennessee; formed part of wider exhibit at the 1984 Summer Olympics in Los Angeles; put on display at environmental education and science centres, zoological gardens, community colleges, National Park visitors centres.

Zambia: used by the Wildlife Conservation Society of Zambia in seminars and leadership training courses for club leaders and teachers.

Science museums and botanical gardens in different parts of the world are using all or parts of the exhibit for display. The Franklin Institute Science Museum in Philadelphia, the Royal Tropical Museum in Amsterdam, the Koenig Museum in Bonn, the Royal Botanic Gardens at Kew, the National Museum of Archeology in Valetta, the Natural History Museums of Santiago, Chile, and Quito, Ecuador, are among these centres, where the MAB panels are enriched by the addition of specimens and other materials of interest to their visiting publics. Elsewhere the Unesco exhibit has been incorporated into much larger events, such as the industrial fair Expoquímica 81 in Barcelona, the bicentennial celebration of the founding of the city of Bangkok, and in conjunction with the World Fair and Olympic Games in the United States.

A host of countries are translating or have translated the exhibit into their own languages: Arabic, Chinese, Dutch, Finnish, German, Greek, Hebrew, Indonesian, Polish, Portuguese, Russian, Sinhala, Swahili, Tamil, Thai and others. Some countries (such as Norway and Italy) have arranged for translated versions of the posters to be reproduced in multiple copies for wider distribution, particularly to schools.

The wide interest in using Ecology in Action as a teaching aid in schools is reflected in an initiative in the Region of Catalonia in Spain, where six posters from the exhibit have been translated into Catalan. Five thousand copies of each have been reproduced by the municipality of Barcelona and distributed to secondary schools in the province accompanied by a teacher's guide. A similar effort has been undertaken in the Basque region of Spain. In some cases, such as the Federal Republic of Germany, the posters are being used not only for science teaching but also for language training.

A Stimulus for New Initiatives

The Ecology in Action exhibit is not expected to meet all the needs of everyone everywhere. This would indeed be impossible given the ecological, cultural and linguistic diversity in different countries throughout the world. In recognition of the inherent limitations of a "universal" approach to communicating ecological information, the exhibit is intended mainly to act as a spur to scientists to take initiatives, if they have not already done so, to communicate their own research results to non-specialist audiences.

To help foster such actions, the exhibit was designed to be used as a core to which other elements can be added: components which address related issues, local examples or research projects, particular audiences, etc. Such additions can take the form of similar posters or panels, audio-visual materials such as films or slide-tape presentations, lectures or discussion sessions, computer simulations, etc.

The fact that the exhibit has been designed in modular form further enhances its adaptability. Each poster can stand on its own, or it can be used as part of a larger module (one of the five sections of the exhibit) or an even larger one (thirty-six posters) to which additional elements can be added. Thus it is possible to apply the core concept not only to the exhibit as a whole, but to each individual poster or each group of posters.

The use of the exhibit as a core, around which new materials on local problems and examples are presented, was well demonstrated at the inaugural six-week showing of Ecology in Action at the Hôtel de Ville in Paris in 1981. On this occasion a selection of about twenty of the Ecology in Action posters was augmented by a French poster exhibit specially produced by the Ministry of the Environment and various French research groups associated with MAB. This series included twenty-five panels dealing specifically with French policy on environmental concerns: control of pollution in air, water, cities; conservation of nature and the cultural heritage; research on tropical forests, arid zones, etc. This exhibition, entitled "La terre entre vos mains" (The World in Your Hands), prepared with a general audience in mind, has subsequently been printed in hundreds of copies and is being distributed to institutions or individuals who may wish to mount their own display.

Special panels based on MAB research at the national level have also been produced in Chile, based on field projects in mountain areas, grazing lands and continental waters. In China, too, environmental panels of special national interest were featured in ten panels created to complement the Unesco posters. These panels covered such problems as desertification control, salinisation, soil erosion, urban modelling and waste utilisation. Efforts for depicting nature conservation were focused on China's three internationally recognised biosphere reserves: Changbaishan in Jilin Province, Wolong in Sichuan, and Dinghushan in Guangdong. The Chinese posters were inspired by local MAB research projects, and their design was the result of close collaboration between scientists, the MAB Committee, and graphic artists attached to the Museum of Natural History in the capital. An estimated 50,000 people visited the inaugural showing of the Chinese-language version of the Ecology in Action exhibit at Beijing Natural History Museum. Wide television coverage, and articles in the *People's Daily* and the local press, as well as in the *Peking Review* which is widely distributed outside China, have also helped rouse the public's interest in the poster materials, which subsequently travelled to most of China's major cities.

The exhibit has also travelled widely in Ireland. Here, a poster competition for schools was organised in conjunction with the exhibit's visit to some ten cities and towns, following its initial display in Dublin in March 1982. The competition was organised by the Irish Art Teachers Association and sponsored by the Irish Heritage Trust, with prizes offered for the contribution of different age groups of schoolchildren on the theme "Ecology—the chain of life".

Inspired in part by Ecology in Action, the Soviet MAB National Committee has produced a set of fifteen posters aimed at schoolchildren aged 11–14. The title of the exhibit is "Environmental Protection" and the full-colour posters address either a single ecosystem type found in the USSR (e.g. tundra, steppe, mixed forest) or particular aspects of land use practices (e.g. effect of pesticides on cereal field community, influence of pollution on aquatic systems) or a specific aspect of nature conservation (e.g. protecting and helping bird life). A teacher's manual supplied with the posters provides detailed guidelines on how to explain each poster and stimulate the children's interest and appreciation of their environment. For example, the teacher might ask: "What is the food web of the agricultural community?" or "What wildlife will be especially affected by crop spraying?"

A final example mentioned here is the work of a group of scientists within the Indonesian Institute of Sciences, who have become increasingly involved in environmental education, as one outlet for scientific information on the Indonesian environment. Pilot schemes have been developed at both formal and non-formal levels. At the formal level, attention has focused on introducing environmental dimensions within the existing school subjects rather than developing separate curricula on environmental issues. At the non-formal level, activities have included writing and drawing contests on environmental issues within children's magazines, contests which attract 700 to 800 entries each (Indonesian MAB National Committee 1984). Some of the winning entries of the drawing and painting contests have been made into picture postcards.

Adaptation of the Posters in Other Forms

The large (120 X 80 cm) poster format of Ecology in Action is appropriate for some purposes, but not for others, and Unesco has received many ideas for adapting all or part of the exhibit. One repeated request has been to produce a book which would "recycle" the posters into another medium for use mostly in schools. It was in response to such requests that the MAB Secretariat developed a joint project with the Committee on the

Teaching of Science (CTS) of ICSU to prepare a series of wall charts which reproduce a selection of the posters in small format (quarter size), accompanied by a booklet comprising complementary new texts, new graphics, classroom exercises and teachers' questions on the subject matter of the various posters selected. The wall charts and accompanying booklet, to be published in English in July 1985 by Heinemann Educational Books of the United Kingdom, are intended for use at the upper secondary school level. They are designed to provide resource material for geography, biology, social and environmental studies students for project and research work, for homework and for general reading. Interest has been expressed so far by other publishers for Spanish and German versions of the book.

Another adaptation has been the presentation of the posters in slide form, for use mostly as a teaching tool. A simple slide package has been produced—in English, French and Spanish versions—consisting of 60 slides (slides of the 36 posters as well as 24 closeups) and a brochure in which the texts of the posters (not always readable on the slides) are reproduced. Transparencies of the posters for use with overhead projectors have also been prepared by the MAB Secretariat, but for internal use only. In effect, cost precludes the reproduction of anything more than a handful of sets in this form, since the pro rata photographic costs involved in the reproduction of transparencies are not greatly reduced with increased production runs.

Conclusion

In both its form and content, Ecology in Action is intended to be an experiment, which, like any other, has both successful and less successful aspects. On the positive side, it can be said that the posters have enjoyed an overwhelmingly favourable response and that they have been considered appropriate for use with diverse interest groups, age levels, institutional contexts and cultural settings. It can also be said that one of the basic aims of this exhibit has been largely fulfilled: namely, to produce something which would serve as a stimulus to those in the MAB family and beyond to take further initiatives to communicate research results to non-specialists. Such initiatives have been many and diverse, including translations into local languages, printing other language versions in hundreds of copies, preparation of new posters and panels, slide programmes, teachers' guides, transparencies.

Accompanying these largely positive indications and reactions have been criticisms which are useful in analysing the shortcomings of the exhibit. It is hoped that for a such as the Bangalore Conference will generate further feedback and critical evaluation of what has been accomplished. Positive and negative criticisms will not only help shape immediate follow-up activities, but will enable this MAB effort to attain one of its objectives—that of contributing to the search for more effective ways of communicating the results of ecological research to lay audiences.

Reference Cited

di Castri F., Hadley M. and Damlamian J. 1982a. Ecology in Action—an exhibit: an experiment in communicating scientific information. *Nature and Resources* **18** (2): 10–17.

di Castri F., Hadley M. and Damlamian J. 1982b. Communicating information on the environment: between emotions and hard data. *Ambio*, **11** (6): 347–354.

Indonesian MAB National Committee. 1984. The scientific community and the development of environmental education in Indonesia. In: F. di Castri, F. W. G. Baker and M. Hadley (Eds.), *Ecology in practice. Vol. 2. The social response*, 236–241. Tycooly, Dublin and Unesco, Paris.

Lefèvre, B. 1984 The ecology of scientific information: telling what we know, asking what we need. In: F. di Castri, F. W. G. Baker and M. Hadley (Eds.), *Ecology in practice*, Vol. 2. *The social response*, 174–182. Tycooly, Dublin and Unesco, Paris.

Unesco. 1978. *Man in the humid tropics.* MAB Audiovisual Series 1. Unesco, Paris.
Unesco. 1979. *Tropical grazing land ecosystems.* A state of knowledge report prepared by Unesco, UNEP and FAO. Natural Resources Research Series 16. Unesco, Paris.
Unesco. 1981. *Ecology in Action: an exhibit. Thumbnail sketches of 36 posters.* Unesco, Paris.
Unesco. 1984a. *Man and mountains/L'homme et la montagne,* MAB Audiovisual Series 2. Unesco, Paris.
Unesco. 1984b. *Man, the key to conservation/L'homme garant de la conservation.* MAB Audiovisual Series 3. Unesco, Paris.
Unesco. 1984c. *Man in arid Lands: nomads in transition/L'homme en zone aride: évolution du nomadisme.* MAB Audiovisual series 4. Unesco, Paris.

It is sometimes thought that, in the age of the video system and the computer, it is no longer worthwhile to use the well-worn techniques of the overhead projector and the slide projector. This is far from the truth and some very imaginative uses for these familiar friends have been developed. One of the recurring themes throughout our workshop sessions was that *all* transfer techniques whether simple or complex have their value and it is more important to select the one most appropriate to the purpose than to worry about whether the technique selected is up to date, or prestigious. It is our students who matter, not the status symbol of our techniques.

Herbert Thier of the Lawrence Hall of Science introduced us to some new ideas. He used an overhead projector to demonstrate just what is meant by "a few parts in a million"—which of course is a familiar phrase in lessons on pollution, etc. He places a series of shallow, clear plastic dishes on the table of the projector and, in advance of the lesson places 45 drops of water in each. He then introduces a dish containing concentrated vegetable colouring, which of course appears intensely coloured on the screen. A medicine dropper is then used to transfer 5 drops of the dye to the first dish, which then, of course has a concentration of one part in ten. Five drops of this are added to the next dish to give one in a hundred, and so on. The children can see the colour getting paler, and there is no colour left by the time they reach one part in a million, and yet they *know* that there is dye present.

He also uses well-photographed slides to transfer research-type evidence to children engaged in a project. For example, a sequence of photographs taken after an earthquake was used to elicit questions about why certain structures were more vulnerable than others, why in some cases an entire building collapsed and in others only part fell down. With carefully chosen sequences a very impressive story can be pieced together where an actual visit to inspect the site would not be possible.

The combination of slides with audio tapes has been brilliantly used by Mark Boulton in his work at the International Centre for Conservation Education in the UK and it is described in the following paper.

AUDIO-VISUAL TECHNIQUES FOR CONSERVATION

MARK N. BOULTON
Director, International Centre for Conservation Education, Greenfield House, Guiting Power, Cheltenham GL54 5TZ, UK

It has become increasingly evident in recent years that progress in the field of conservation and development is, to a large degree, dependent on an informed and sympathetic public. Promoting greater understanding of conservation and the environment should therefore be an essential ingredient of all conservation programmes. Whilst modern science continues to develop a bewildering variety of new technology for the would-be communicator, the purpose of this paper is to underline the important role that even the more modest audio-visual techniques can play in conservation education.

There is of course nothing new in using colour slides to illustrate lectures and talks—whether on conservation, climbing, or cookery! In the last decade however, audio-visual techniques have evolved with astonishing rapidity. Audio-visual "spectaculars" involving dozens of slide projectors often interlinked with 16 mm cine projectors and controlled by computers are becoming commonplace at large exhibitions and conferences.

Although strictly speaking, any presentation which links images with the spoken word can be termed "audio-visual"—from the live use of a blackboard or flip chart to the present day video disc and three-dimensional holography, this paper will confine itself to the relevance of filmstrips and slide presentations to the field of conservation.

Why Audio-visual?

If as the Chinese proverb suggests, "a picture is worth a thousand words", then a series of dramatic visuals accompanied by a carefully narrated text should prove to be an extremely effective way of spreading the conservation message. Research on the efficiency of information transfer suggests that the average student remembers less than 10 % of what he reads, about 20 % of what he hears and about 30 % of what he sees. However, the combination of sight and sound together increases retention to more than 50 % , and subsequent discussion can raise this above 70 % . Since audio-visual by definition utilises both sound and sight—usually simultaneously, then its potential as a tool for communicating conservation is clearly evident. Having made the decision to use AV techniques however, the potential user is left with a mass of unanswered questions on the different methods of presentation, the nature and cost of the equipment, whether to "buy-in" prepared programmes or prepare them oneself, and where to obtain further advice and information.

Why Filmstrips and Slide Programmes?

Why use filmstrips and slide programmes in preference to films and video? Whilst all these media have their own contribution to make in the conservation field, filmstrips and slides do offer several distinct and in some cases, unique advantages.

Economy

Both are relatively cheap to purchase and to produce. It is generally reckoned that the production of a slide–tape programme costs less than 5 % of a comparable 16 mm film and copies of a pack of 60 slides may cost less than UK£10 each compared with the average 16 mm film at UK£300 or more.

Flexibility

A programme purchased as a 60 frame filmstrip for around UK£5 can be used unaltered with the script provided, or may be cut into separate frames and mounted to make slides. These in turn may be rearranged to suit the "live" message of the presenter or mixed with other resource materials as required. Programmes are also readily updated by substitution or addition of new material.

Quality

There is a widely held view that filmstrips and slide duplicates are necessarily "inferior" to the more sophisticated film and video materials. In fact, the reverse should be the case! Original slides (and good quality copies too) may be enlarged using a good projector to 4 metres or more across, and the sound quality from a well-prepared cassette far exceeds that of a 16 mm film optical soundtrack. Mix slide and tape well with a good twin projector "cross-fade" system and the audience is never quite sure if it has been watching a film or a slide show—but is usually very impressed with the presentation.

Speed

The production time for filmstrip and slide presentations is much shorter than for 16 mm films (which are rarely completed within a year). Trainees at the International Centre for Conservation Education (ICCE) have completed slide–tape programmes within a fortnight!

Filmstrips or Slide Programmes?

Although it is possible to originate AV programmes so that they can be distributed as both filmstrips *and* slide-packs it is appropriate to consider the relative advantages of each.

Most of the early AV materials—certainly those used in schools, were supplied as Filmstrips—continuous lengths of 35 mm film comprising a series of images or frames. Some of these were "full-frame" (image size 36 X 24 mm) others were "half-frame" (image size 24 X 18 mm). Such filmstrips were cheap to produce, light to distribute (especially overseas) and also eliminated the risk of losing individual frames—always a danger with slide-packs. Once the user established how to thread the filmstrip correctly into the projector (some readers will know from bitter experience that even the humble filmstrip can be loaded in four different ways, only one of which projects a picture the correct way up and the right way round!) the sequence of appearance of the pictures is fixed and the frames cannot be projected in the wrong order. However, early filmstrips were often of very poor quality and it was sometimes difficult to tell whether they were colour or black and white! Fortunately there have been substantial improvements in quality in recent years. Some found the inability to change the sequence of pictures a serious disadvantage, and unless fairly sophisticated production facilities are to hand, updating of filmstrips is impossible. In general, filmstrips are also unsuitable for use with pre-recorded tape/cassette commentaries although there are a few manufacturers who market a filmstrip projector which incorporates a cassette recorder. Such units automatically advance the filmstrip one frame at a time in response to signals received from the "synch" track of the cassette. Perhaps the greatest advantage of filmstrips is their low cost—about half that of a comparable slide-pack. Those on economy budgets may consider it worth purchasing AV programmes in filmstrip form and then cutting and mounting the individual frames into standard slide mounts. Before doing so it is wise to check that a "full-frame" filmstrip is being purchased and not a "half-frame" version.

Slide-packs would now appear to be the more popular way of presenting AV programmes. Most colour slide films are returned from the processing laboratories in "ready to project" card or plastic mounts, and a wide variety of 35 mm slide projectors are now available to suit most pockets. Relatively basic projectors utilising straight trays or magazines into which 36 or more slides are pre-loaded can be purchased for around UK£40 and some of them include remote control facilities. Rotary magazine projectors such as the well-known Kodak Carousel, which carry 80 (occasionally 140) slides have become the accepted standard for "professional" presentations though at the top of their range these will cost UK£500. Such projectors have many additional facilities including built-in spare bulbs, exchangeable lenses, thermal cut-outs to prevent damage by overheating, and remote sockets allowing sophisticated electronic control and coupling with other projectors and even computers for multi-image/multi-screen presentations. Fortunately even some of the cheaper projectors now carry the "DIN" socket which is necessary if the projector is to be coupled with a tape/cassette recorder for automatic presentations.

A universal requirement for both filmstrip and slide projectors is of course a source of electricity. Some projectors can be adapted to operate from 12 volt car batteries though they draw heavily on the current and the batteries must be recharged very frequently. Where mains electricity is not available all will operate perfectly well from even the smallest petrol generators. (The World Wildlife Fund is using dozens of these for rural conservation education programmes in developing countries.)

The "Live" Presentation

The potential of the "live" presentation should never be underestimated—though its final impact will depend to a great extent on the skill and knowledge of the presenter. Thorough preparation is essential and slides should be chosen not merely for their technical and aesthetic value (though both are important) but so that they form a logical series of visual images which help to guide the theme of the talk. Time spent preparing title slides, occasional graphics and end credits make for a more professional approach. Whenever possible the projector should be situated at the rear of the projection room (rather than in the middle of the audience) and it may often be necessary to use a long focus lens (180 or 250 mm). The projector should ideally be controlled by the presenter and this is usually accomplished using a remote hand control on a long extension lead (although infra-red cordless controls are becoming more readily available). Care should be taken to avoid talking too directly about the projected slides (the "this as you will see is a tiger in long grass" approach!) since this undervalues the intelligence of the audience. The most effective presentations are often those where the visuals form a "structured background" to a talk which would be interesting in its own right—even without the pictures.

Slide—Tape Presentations

There will always be those who for reasons of time, or lack of specialised knowledge or confidence, prefer to utilise pre-recorded narrations. Although these are less personal and "direct" than a live presenter, they have the advantages of a professional voice, music and/or sound effects, and of being predictable in length (unlike some "live" presenters!). Whilst it is possible to use such slide—tapes packages by playing the cassette on a domestic cassette recorder (assuming the sound output is adequate) and advancing the slides manually, this requires very thorough preparation and complete concentration during the performance. Those considering the regular use of such presentations would be well advised to consider purchasing special slide—tape equipment.

Although a number of "magic boxes" are advertised which can, in theory, link any slide projector to any cassette/tape recorder, the serious user will need access to a cassette recorder having an "AV head". Although such units still use the standard international compact cassette, they are specially constructed to respond to control .pulses (or "cue-tones") carried on track 4 of the cassette tape, whilst the audio (narration/music/sounds effects) is carried on tracks 1 and 2. Track 3 is usually left blank to avoid any possible interference between the "inaudible" control pulses and the commentary (though on some recent formats tracks 3 and 4 are used to increase the reliability of detection of the pulses). It follows of course that AV cassettes can only be used "one-way"—they cannot be turned over and used on the other side like domestic cassettes. It also follows that if one is confronted with a troublesome cassette which does not appear to operate correctly with a slide–tape unit, it can be turned over and played "upside down" to find out if the control pulses are present. A number of manufacturers make basic units which apart from their AV head, differ little from ordinary domestic recorders. Examples are the Philips AV cassette, and the Hanimex synchrocorder. Much better, however, are units made especially with slide–tape in mind—such as the Coomber 353 AV machine (around UK£225 in UK) which combines a high quality amplifier and public address system (with large built-in speaker) with the AV cassette unit. ICCE has designed a special "Pandamatic" system for use under tropical field conditions where the electronics are protected against dust and adverse climate by enclosure in an extremely durable rigidised aluminium case. With such equipment, all that is necessary to present a pre-recorded slide–tape show is to load the slide magazine on to an appropriate projector, link it to the "Pandamatic" with a single lead, place the cassette in the AV cassette deck, and press the play button. The pre-recorded (1kHz) pulses automatically change the slides at the correct places in the narration.

More sophisticated units can control two or more projectors—not merely advancing the slides automatically, but fading or dissolving from one image to the next—thus avoiding the blank/black screen interval which occurs in most single projector presentations. Is such an expensive technique justified? (this system costs at least double that of an ordinary single slide–tape unit). The answer depends very much on the intended use, the level of sophistication of the audience, the budget available, and the experience of the operator. Such units are ideal for permanent installations in visitor centres, for major lecture tours, and other special presentations—but in a "third-world" situation it would almost certainly be preferable (budget permitting) to purchase two single AV units, and to use them in different locations. There is little doubt, however, that the two projector system is an extremely effective medium, whether used "live" or with pre-recorded cassettes. It allows for very professional presentations with the possibility of a wide range of special effects such as the building up of complex graphics or the changing of one scene dramatically into another. As a general guideline, the total cost of two Kodak Carousel projectors, long-focus lenses, and a portable AV/amplifier/cross-fade system would be about the same as the cost of a 16 mm projector but the possibilities for the creative user are virtually limitless. Finally, it is also worth noting that twin-projector "cross-fade" presentations may be very effectively transferred to "video" although access to specialised equipment is required to obtain the best results.

To Purchase or to Produce?

An increasing range of good AV presentations on conservation issues—both national and international is becoming available. Programmes on threatened species and habitat loss, as well as the broader issues of population, pollution and diminishing resources can all be found in Educational/AV sales catalogues with prices ranging from as low as UK£3 for the small resource packs to UK£25 for longer presentations complete with cassettes. Although it may be possible to obtain such programmes on approval, producers are often somewhat reluctant to risk the possibility of damage by poor handling. A number of these presentations are reviewed in the conservation press and it is usually possible to obtain a good idea of the content and quality of the materials from such reports.

Sooner or later, however, one may wish to deal with a topic on which no AV material seems to exist, or to use an approach which differs from the available presentations. How does one go about producing an AV programme? What skills are needed? How much will it cost? Limitations of space only allow mention of a few general guidelines on the subject. Perhaps the most basic question is, "Which comes first—the text or the pictures?". Experience would suggest that many "live" lectures on conservation are in fact based on the available slides, the story-line being developed to suit the pictures in hand. Providing enough material is available and the presenter well-informed, such presentations are often very effective and could well form the basis of an "AV" programme. Several major presentations prepared by the author originated in this manner. On the whole, however, it is better to prepare a "story-line" first and then, on the basis of the slide material available, develop the story-board into a full script. Rarely is it possible to locate all the "ideal" visuals but at the same time, excellent slides turn up which were not anticipated. Modifications may then need to be made to the text to accommodate them. The soundest procedure is perhaps to first prepare a written response to the following questions:

- what is the major objective/message of the proposed AV?
- what is the target audience? (old/young; educated/illiterate?)
- will it be used live (by reading from a prepared script) or using a pre-recorded narration? (with music/sound effects?)
- are the necessary original colour slides available? If not, will others be prepared to loan their material or will it be necessary to obtain them from a photo-agency? (and adjust the budget accordingly!)
- are the required skills available to prepare the script, select the slides, undertake the graphics, record the narration, "mix" in the music and sound effects, or will some (perhaps all) of these tasks need to be delegated to others? (and how much will this cost?)
- how many copies are required of the finished presentation?

Some impressive AVs have been prepared by a single individual, but normally substantial additional expertise will be required if the programme is to be fully effective. The answers to the above questions will help in programme planning, to estimate the production costs, and to seek technical assistance/sponsorship as necessary. It is impossible to specify the exact budget required since there are so many variables. One of the most effective and popular presentations prepared by the former WWF/IUCN International Education Project ("Why conserve wildlife?") was originated for less than UK£50. The use of all the slides, scripting, graphics and the narration by Sir Peter Scott were all donated and the sound mixing was carried out on a domestic stereo tape recorder. In contrast, the more sophisticated presentations prepared recently for twin projectors have had to carry the costs of "bought-in" slides, hired studios, paid narrators, and music copyright fees. It is of course possible to put the whole presentation out to a commercial AV production house, but find your sponsor first! If the decision is taken to tackle the production personally then the most important requirements are probably motivation, the capacity to learn new skills and sheer dogged perseverance! And in case one gets too wrapped up in the technical intricacies of the process, it is wise to get some constructive criticism from others at appropriate stages in the production.

ICCE has, in many instances been involved in co-productions with other organisations. Programmes as varied as "Turtles in Danger" (with the Species Survival Commission of IUCN), "Underwater Conservation Code" (with the Marine Conservation

Society), "Focus on Otters" (with the Royal Society for Nature Conservation), "Renewable Energy for today and tomorrow" (with the National Centre of Alternative Technology), "Desertification—its causes and some solutions" (with UNEP)—and "Planning for Survival" (with the Education Commission of IUCN) were all produced in this manner.

Are AVs Effective?

How does one measure success? By the size of the audience, or by the "oohs and aahs" after the performance? By the number of programmes sold or by the solving of conservation problems? Almost 10 years of involvement in AV work suggests that all of the above can be positive indicators. AVs can certainly increase the level of awareness on conservation issues; they can suggest specific action required to tackle conservation problems; they can "motivate" and "enthuse" those who watch them. More than 1000 packs of just one AV programme "Saving the whale" have now been distributed to a dozen or more countries and an Icelandic version has been prepared by whale conservation groups for a campaign in that country. "The Mountain Gorilla"—prepared to support the Mountain Gorilla Project in Rwanda and narrated in both Rwandese and French, has made a major contribution to raising the level of environmental consciousness in that country. More than 300 copies of the programme prepared to support the WWF Tropical Forest and Primates Campaign ("A green earth or a dry desert?") have been widely distributed in English, French and Spanish. A "multi-media" pack for children based on the World Conservation Strategy is in widespread use in British Schools and has been adapted for use in Spain and Belgium (where 6000 copies were distributed free by one of the large banks). The indications are certainly very positive.

But a final note of caution. AVs are not the "magic answer" to all the world's conservation problems. They will not prevent the extinction of species. They will not solve the problem of acid rain. They will not, by themselves result in a greater use of renewable energy or save the rain forest. What they can, indeed will do, however, is to promote greater awareness of conservation and the environment, and increase the level of knowledge and concern. After all, this is the vital prerequisite for informed debate and appropriate action.

Obviously all these techniques are not equally suited to all purposes, nor are they equally available to all countries. We continually come back to the key problem—the teacher. In the hands of a skilled and imaginative teacher the simplest of materials can be made to convey a powerful message. And in the hands of a poor teacher, even the most expensive and sophisticated of high-technology developments can be counter productive in terms of the information transferred. The techniques discussed in this chapter should be regarded as a menu of ideas, some of which may be copied, some of which may be a trigger for quite different adaptations. None of them is a guaranteed formula for success without dedicated work on the part of the teachers and others who will use them.

4

Networks—Nets that Work

PETER TOWSE

A NETWORK is merely a system of information transfer linking one person with a particular piece of information to another in search of it. A network is thus any means by which we are able to learn new concepts, new ideas, new experiences, new practices—in short, to discover new solutions to old problems. This exchange can be as complicated or as simple as the circumstances dictate.

Robert Silber has provided the following overview of networking:

> "Oo-ooo, I heard it on the grapevine."
> Gladys Knight and the Pips

Networking is quite simply people talking to each other, sharing information and resources. The "grapevine" has always been an important source of information and support in both personal and professional life. The arrival of the information age has heightened the significance of networking in our professional activities. Technology has opened new channels of communication, while the rapid pace of change and a flood of new information have increased the value of direct contact with others with similar interests.

The term networking is much in vogue. It can refer to anything from a computerised information service to a group of friends. In this overview, we emphasise the strengths of resource exchange networks, which are based on personal contact and mutual exchange.

Resource exchange networks exist to share information, ideas, materials, expertise, personal services, and other resources. They cross over the usual categories that organise our work—the institution and the profession—and make new connections. Any group of people with a common interest can form a network by making the effort to stay in touch and actively share resources and information. When resources are limited, as they always are, networks can help find what we need. When we are flooded with information, networks can help filter out the information and ideas that are worthwhile.

The power of networking comes from the fact that networks are based on direct personal contact and mutual exchange. The most influential source of information about new ideas and practices is our peers. Studies of educational change suggest that personal contact is critical for acceptance of an innovation. Our everyday experience supports this conclusion; we trust information that comes from people we know. In addition, personal networks provide psychological support as well as information and resources. We can share our successes or vent our frustrations and get a helpful response—sometimes an idea or resource and sometimes just a friendly ear.

Effective networking often requires a fresh look at the resources we have and those we need. Ideas, expertise, energy, and information are resources as much as materials and money. We need to take a broad view of ourselves and others as human resources. The confines of a profession or organisation often lead us to define our resources too narrowly. Special educators turn to other educators for advice and university faculties go to other professors when, perhaps, a person from another field could provide a fresh perspective that would be more useful. Networks can cross professional and organisational lines and offer the personal contact needed for such flexible exchanges.

Networks also depend on key people who co-ordinate information and help put members in touch. The skill of a co-ordinator is to see the connections between the needs of one person and the resources of another.

Some networks exist primarily to share information. Direct contact is a fast and efficient means for such exchange. It is often possible to get the answer to a question faster with a few phone calls than by checking the literature. In rapidly changing fields, direct contact is often the only way to get current information. More important, a question can lead to a problem solving discussion that may produce unexpected results.

Resource exchange networks can be formal or informal. Usually, they begin informally at the grassroots level as a group of people with a strong common interest or an action cause. Many networks remain informal and small, while others grow more formal, acquiring a newsletter and an organisation. Despite growth, many networks, such as the parent groups mentioned above, have maintained their emphasis on direct personal contact among members.

Leo Tan gave an account of networking at its best at the school level when he described how the Singapore Science Centre provides innovative, motivating and stimulating ways to convey science and technology information through such media as participatory exhibits, posters, live specimens, panoramic displays, slide shows, microcomputers, video presentations—or appropriate combinations of some or all of these. For the learning of science to be of any real value, the subject matter has to relate to the everyday lives of the pupils and teachers and science centres are particularly well placed to show this relevance in an educational and entertaining manner. Because they are not bound by syllabus constraints they can thus complement the more formal work in science covered in schools, helping to consolidate various science concepts, stimulate scientific curiosity and cultivate creative thinking. In so doing, they are able to encourage the development of desirable scientific attitudes and skills, and facilitate further learning in science. A further advantage of their non-formality is their capacity to encourage links between professionals in education, industry and the media, links which will result in the production of the very finest educational materials, be they in the form of slides, films, books, journals, posters or exhibitions. How successful his own centre has been in achieving these aims may be seen from the examples provided in his paper "The role of the Singapore Science Centre in science and technology education":

Science Centres throughout the world are playing an important if not essential role in meeting the educational needs of society. They are being established wherever the actual and potential benefits of these institutions in creating a greater appreciation and understanding of science and technology among the general public and the student population are recognised.

The Singapore Science Centre is a major institution for the dissemination and popularisation of science and technology to the students and the general public in Singapore. It is a Statutory Board under the Ministry of Education.

Its objectives are:

(a) to exhibit objects illustrative of the physical sciences, life sciences, applied sciences, technology and industry and

(b) to promote the dissemination of knowledge in science and technology.

Since its official opening in December 1977, the Centre has attracted well over two and a quarter million visitors with students constituting about 60% of the total. The numbers are increasing. In the three preceding Financial Years 1982/83 to 1984/85, there were 347,984, 410,085 and 463,063 visitors respectively. In three visitor surveys since 1982, 70–80% of the visitors indicated they learned and enjoyed themselves and would make revisits, while a mere 2–2½% felt their visit was unproductive. More than 95% of the schools in Singapore participate in the Centre's Institutional Membership Scheme which encourages schools to arrange visits to the Science Centre by their students for educational tours and programmes. In FY 1984/85, more than 176,000 students took part in the Centre's education and exhibition programmes under the scheme.

Members of the public and youth are also encouraged to be "Friends of the Science Centre", a scheme which entitles a member among other benefits, to a one year complimentary pass to the Centre, free subscriptions to the Centre's quarterly magazine and newsletter, access to the Reference Library, and participation in activities organised by the Centre. In 1984/85, there were almost 8,000 "Friends". This is nearly double the number in 1982/83 and more than 26 times that in 1978/79 when the scheme was initiated.

The above figures serve to indicate that the Science Centre has been playing an increasingly important role as a non-formal educational institution. For the students in particular, this means there is a complementary system to the formal science curriculum that will help reinforce and exemplify concepts, cultivate creative thinking and arouse scientific curiosity. In so doing, desirable scientific attitudes and skills can be developed which will enhance and facilitate further learning in science.

The Science Centre is thus seen as a centre of learning where self-motivating experiences in learning through the use of the senses and the intellect are encouraged. Instead of a "hands-off" policy, visitors are encouraged to touch, manipulate and become actively and pleasurably involved with the exhibits. By interacting with the exhibits, the visitor is able to learn science through exploration and discovery.

The Singapore Science Centre like its counterparts in other countries, is less constrained than the formal school-based science system in the types of programmes and facilities it can offer. As a result of this, its programmes tend to be innovative and stimulating.

The Singapore Science Centre performs its role in science and technology education for both students and the community at large by providing specialised facilities which include the following:

(a) Exhibition Programmes

(b) Science Education Programmes for Schools

(c) Science Publications Programme

(d) Science Promotion Programmes

The objectives of the Centre are met through these spheres of activities. While the overall goal is to motivate and stimulate the visitors to learn or take an interest in science and technology and their relevance to everyday life and industry, the major thrust must be directed at the students from pre-Schools to pre-University levels.

Each of the above programmes is briefly summarised below.

Exhibition Programmes

The general philosophy which is followed in the choice of themes and the approach for the setting up of the exhibits is that the themes are pertinent to educational, economic or technological developments in Singapore and the exhibits explain basic concepts of science and their applications to daily life and industry. In this way, the student and the layperson can relate science and technology to their everyday lives and the society they form part of.

The Science Centre has some 6,500 square metres of space devoted to exhibitions on pertinent or contemporary themes and issues (e.g. Electronics (including Computers), Ecology, Energy, Genetics, Aviation, Population, Mathematics etc). In all, about 500 participatory exhibits are found in five exhibition halls viz, the Solar Lobby, the Physical Sciences Gallery, the Life Sciences Gallery, the Aviation Gallery and the newly-opened Discovery Centre.

The scientific concepts are conveyed in a simple and systematic way, through participatory exhibits, graphics, live specimens, dioramas, microcomputers, multi-media presentations and selected video shows.

Three criteria have to be met in the presentation of the exhibits. Firstly, the scientific content must be accurate. The assistance of scientists and other professionals is enlisted for every new project. Secondly, the exhibits have to be visually aesthetic and inviting. They should tease, entice, excite, provoke thought and motivate the visitor. Thirdly, two factors among others have to be considered during exhibit construction, i.e. visitor safety and the manhandling of exhibits by visitors.

Thus research and development into exhibition design and fabrication (taking into account visitor expectation and behaviour) are imperative. Those who have followed the development of the Centre's exhibition evolution through the years will have noticed the gradual but practical change from solid wall presentations to modular forms and the growing emphasis on the "hands-on" approach. More than 60% of the exhibits involve direct, active participation. Visitors can do one or more of several actions at each exhibit besides musing. They can touch, step on, turn cranks, push buttons, trace patterns, clap, stomp, operate computer keyboards, and even taste and smell. Through the use of their senses, visitors are able to find out for themselves how scientific principles work.

Almost all the exhibition themes in the various galleries are related to the science syllabi of the schools and the exhibits serve as excellent material for complementing science teaching in primary and secondary schools and junior colleges.

The Science Centre updates and revises about 10% of its exhibits yearly. There are two kinds of exhibitions—long term (i.e. with lifespans measured in years but subject to periodic updating) and temporary, each lasting a few days to 6 months to keep students and the public abreast of the latest scientific information.

Guided exhibit tours for organised groups are provided from Tuesdays to Sundays. These tours have to be booked in advance and worksheets based on the exhibits are available upon request. The purpose of the worksheets is to enable the visitor to evaluate their knowledge gained from the exhibits and also serve as a feedback to the Centre for possible follow-up action.

Exhibit workshops for teachers are also organised to familiarise teachers with the facilities of the Centre and to impress on them the importance of the Centre as a resource centre. Teachers can thus make use of these exhibits to reinforce their classroom lessons as well as challenge the young to see beyond the textbooks and curricula and thus put them in good stead for the real world conditions they would experience upon graduation.

Science Education Programmes for Schools

The Science Centre runs a series of active, educational programmes for students (up to Pre-University level) and teachers throughout the year. The programmes are specially designed to complement and enrich the school science syllabus and cater to the needs of pupils of different levels. They also provide facilities and specialised equipment (e.g. ecology laboratory, lasers) which schools do not or are unable to provide. The programmes also enable the students to be aware of the latest advances in science and technology. Another priority of the Centre's school programmes is to lower the student/teacher ratio. This of course raises a problem in that increasing the number of activities (as the Centre is doing) does not guarantee a corresponding increase in student numbers. The total number of students attending only school programmes in FY 1983/84 and FY 1984/85 was 68,279 and 75,571 respectively. The numbers are near saturation with the present manpower strength. However, the Centre is optimistic this hurdle can be overcome.

The lowering of the student/teacher ratio permits an approach which is activity-oriented and pupil-centred, relying on first-hand observation and direct experience outside the schoolroom. The programmes include laboratory courses and demonstrations; science talks, shows, forums and seminars; filmshows and activity classes. It would be pertinent to note that school principals and teachers are consulted on the choice of topics to ensure the programmes complement and not duplicate school efforts. Teachers have workshops, talks, forums and seminars arranged for them, usually on Saturdays and during the school holidays. Feedback from teachers and students is a prerequisite for all the Centre's educational programmes. The Centre is also awaiting the results of an independent research study on its secondary school programmes by a graduate teacher (M. Ed dissertation).

Although there are nine laboratories (excluding the 10,000 square metre Ecogarden and Ecolab) in the Science Centre, there are space and manpower constraints which tend to limit the rate of growth of the school programmes. The shortage of teaching space has necessitated the partitioning of two of the Exhibition areas into a Mathematics Activity Room and a Computer Corner.

All the school programmes (which have to be reserved on a first-come, first-served basis) are fully booked very soon after the beginning of each term when reservations open.

Among the recent additions to the educational facilities of the Centre are:

(a) the ECOLAB to help in the teaching of ecology by providing an outdoor laboratory equipped with essential apparatus and instruments for collecting and examining plants and animals obtained in the ECOGARDEN.

(b) the PRIMARY SCIENCE ROOM to support the teaching of science in the primary schools and to serve as an operating model for schools to adopt. It is fully equipped. At present, only a third of the Republic's primary schools have Science Rooms.

(c) The COMPUTER CORNER with 15 microcomputers in network, to provide students, teachers and "Friends" a better understanding of computers by using them to develop their own programs and encourage teachers to use the microcomputer as an effective tool in the teaching of science and mathematics.

(d) the ACTIVITY ROOM to encourage the learning of science in lower primary students (Pr 1-3) and pre-Schoolers through a play-oriented curriculum.

The Science Centre works in collaboration with the Ministry of Education, Science Teachers Association of Singapore and the Curriculum Development Institute of Singapore for the development of many of the above facilities and teachers' workshops.

There are also two teachers-on-attachment at present. They are involved in developing teaching modules for the Primary Science Room and conducting workshops for primary science and primary and secondary ecology teachers.

Science Publications Programme

To achieve its objective of enriching the knowledge of the student and the general public in science and technology, the Centre produces and publishes the following:

(a) SINGAPORE SCIENTIST, a quarterly magazine containing news and articles on current events or issues in science and technology. There is also a special section for students incorporating experiments, quizzes and special features to supplement science learning. This is the most popular science magazine in Singapore with a circulation of about 30,000 per issue and a readership exceeding 100,000.

(b) WALLCHARTS which are colourful and illustrative broadsheets on various scientific topics. They contain explanations, information, experiments, puzzles, etc., on important science concepts. They are meant as teaching aids in schools but students and parents would find them equally beneficial. A total of 23 broadsheets have been produced. They complement the wallcharts produced by the Curriculum Development Institute of Singapore.

(c) NATURAL HISTORY GUIDE BOOKS on animals and plants in Singapore (but applicable to the region). The publication of this series is in response to the paucity of educational materials on local flora and fauna. These fully illustrated and colourful books are written by competent authors and are intended for a very wide audience—from the serious-minded science student to the casual nature lover.

(d) 35 mm SLIDE PACKAGES to accompany the guide books. These are being produced in response to requests from teachers for more teaching aids and audio-visual software.

(e) EXHIBITION HANDBOOKS to enable interested students, teachers and parents to delve deeper into topics on display in the Exhibition galleries. A series of Mathematics handbooks has been prepared to accompany the MATHEMATICA EXHIBITION.

(f) SCIENCE CENTRE NEWS, a quarterly newsletter distributed to all schools and ''Friends''. It contains information on the activities and publications of the Centre and recent or forthcoming additions to the Exhibition and Education programmes/facilities.

Science Promotion Programmes

The Science Centre collaborates with the Science Teachers Association of Singapore, the Singapore Association for the Advancement of Science, the Singapore National Academy of Science, tertiary institutions and other scientific bodies in organising innovative activities to promote science education in Singapore. Some of these are regular events viz. the annual Singapore Youth Science Fortnight (now in its 8th year) for school students and teachers; the biennial Technology Fair for tertiary students; the biennial Ecology Week alternating with the biennial Energy Week for students, teachers and the public; and the increasingly popular Primary Science Club Activities scheme for upper primary students.

Public and industry-oriented talks and seminars, competitions, etc., are held frequently to enable both students and laypersons to appreciate the relevance of science and technology to our society. The recently completed Science Video Fair

Competition and Exhibition brought home the fact that there is abundant talent (and facilities) in schools, tertiary institutions, offices and homes, waiting to be tapped to produce innovative and educational video recordings of local scientific interest.

What Next?

What else can the Science Centre do for science education besides continuing to introduce new programmes and innovative activities to promote science education through the unique concept of "hands-on" and discovery?

Singapore is moving towards more advanced technology to raise productivity so as to sustain and improve its social and economic progress. It requires youth and the adult population to be educated and informed of the latest developments in science and technology (e.g. biotechnology, computers, lasers, fibre-optics, etc.) in order to meet the challenges of the technological era.

High technology exhibitions will be developed under the existing Science Centre programmes but Singaporeans can look forward to 1987 when a new education facility—the OMNIPLANETARIUM (the first of its kind in South-East Asia) and its accompanying science and technology galleries (all housed in a new building complex) will be ready to educate and entertain them in a way that they have never experienced before.

The Omniplanetarium uses the latest audio-visual technology for communicating science and technology. It has two dome-screen projector systems, one a planetarium projector and the other a unique film-projection system called the Omnimax System. The dome screen is tilted and planetarium programmes in combination with Omnimax films will enable the teaching not only of astronomy but many topics in geography, chemistry, physics, mathematics, etc., can be enhanced through the use of the planetarium.

Within the planetarium environment, students can develop useful mental skills, starting with simpler ones such as observing and categorising. These skills are combined with further skills such as analysing and interpreting. At the top of the mental skills hierarchy is problem solving. These skills are the prerequisites in the learning of science and acquiring of technological skills.

Conclusion

In concluding, it must be recognised that no single organisation can achieve everything by itself and hence the Singapore Science Centre together with the schools and scientific organisation/institutions will seek to find more ways to work together for the improvement of science education in Singapore and hopefully for the identification of a "science culture" in our society.

Similarly, Nancy Law described how the Hong Kong Association for Science and Mathematics functions as a communications network on various levels. The Association produces a number of publications, not only a half-yearly journal and a monthly newsletter but also a number of subject bulletins and feature booklets on certain topics which are produced on an irregular basis and distributed by request. Activities such as workshops, seminars, field trips, visits, etc., involving as they do a two-way system of communication, prove more effective than journals, although they obviously only involve the more enthusiastic and motivated teachers who volunteer for such activities. Then again, the Association promotes new course materials and teaching approaches by producing tape-slide packages, audio-visual materials and low-cost equipment.

But perhaps its greatest impact is in the courses it provides for teachers. Through orientation courses it establishes a line of communication with beginning teachers and it is also responsible for a programme of school-based in-service education for teachers, known by the acronym INSET. The pilot work for INSET was initiated in three schools by enthusiastic and experienced teachers at the colleague, departmental and school level and has engaged teachers in the improvement of science teaching. For example, some schools have been at pains to organise the timetabling in such a way that teachers in the same department have a double period free at the same time, so that they can hold seminars, workshops, demonstrations, etc., among themselves, each teacher contributing his own special skills. The more experienced teachers may contribute more in the way of such skills as laboratory safety and classroom management, while their younger colleagues may contribute more in the way of new areas in the curriculum, such as electronics. In addition, the programme encourages the teachers to observe each other teach and engage in various forms of team teaching. An important point here is that science teaching should also involve those from outside the subject discipline. For example, in a discussion of the recent disaster at Bhopal, a social studies teacher may be able to add a new dimension to the problems of chemical pollution.

Jorge Barojas has emphasised the importance of communication as a strategy for development and pithily defined a network as "just a net that works". In Mexico, his and his colleagues' efforts to maximise the exchange of new ideas and concepts have been particularly channelled into the exchange of information through journals, books, exhibits, teacher training, educational research and an understanding of the role of women (this last through a project with the acronym MACHO!). A particularly interesting example of their educational network in practice is the compilation of Spanish textbooks, for example on modern physics. To have as textbooks only translations of books produced outside the country is a "measure of underdevelopment". Such translations, of course, are essential to add an international dimension and to make available to students work that would otherwise be outside their reach, but these should be supplemented by books which have been produced within the country. In Mexico, a number of university physics students have been involved in the production of secondary textbooks, for example those in modern physics.

Among the many activities used to popularise science and technology in China are hobby activities, science fairs, forums, contests and computer programming. Also, under the "Love Science Month", children throughout the country are encouraged to read a popular science book, carry out a small scientific experiment, make a scientific or technical appliance, observe and record some natural phenomenon, acquaint themselves with the achievements of a particular scientist, and attend a popular science lecture—all in the small space of a month!

In developing countries more than in developed ones, information networks tend to concentrate more on the secondary sector than on any other. Meanwhile, as Elore Alonge of Nigeria has pointed out, little or nothing is done to establish effective network systems in the non-formal sector, even though farmers, traders, etc., usually constitute the largest section of the population in

FIG. 4.1. A montage of some of the posters on display.

FIG. 4.2. Picture from a Science Centre publicity document from China.

a developing country. In spite of the fact that "science and technology for future human needs naturally involves every age group, literacy level and level of decision-making", the neglect of this non-formal sector was highlighted by reference to such events as the large number of "kerosene blasts" recently in Nigeria, as well as to more universal events such as the multiplication of pesticide levels in food chains, the pollution effects of fertiliser on untested soils and crops, etc. Such a wide range of events emphasises the need to identify the appropriate level of sophistication for each particular network.

Some of these experiences are reinforced by Muhammad Ibrahim's description of the work of Bangladesh's Centre for Mass Education in Science in Dacca:

1. Introduction

The Centre for Mass Education in Science is a voluntary organisation with an aim to bring science and technology nearer to the general people of Bangladesh through a mass education at the grass root level.

The Centre is registered with the Registrar of Joint Stock Companies under the Societies Act of 1860. This is also registered with the Social Welfare Department and External Resources Division as an organisation receiving foreign donation for the purpose of undertaking voluntary activities in Bangladesh.

The following are some of the activities it has undertaken in recent years. These include innovative works on education, the publication of books and magazines, an experimental rural technical school for the disadvantaged children, and the creation of an unconventional TV series on mass literacy.

2. Innovative Works

The Centre for Mass Education in Science has undertaken innovative works and experimentations in the field of reading materials and teaching aids for both formal and non-formal education. Much of these have been published in its associated periodical monthly Bijnan Samoeeki.

As an example, a proposal by the centre to use an extra label to the match box with letters and syllable printed on it, has been received with enthusiasm. The proposal published in Bijnan Samoeeki (March 1979) in all details, was presented in a workshop on Mass literacy and Rural Development. This could provide a very cheap and automatically distributed gadget for Mass literacy. Bangladesh Association for Community Education and the Chemical Industries Corporation later actually made an experimental trial of the gadget, while the centre helped in the design of the project. Some innovative ideas were experimented with through the associate Science Club of the Centre, the Anushandhani, of which there are over fifty branches throughout the country. Trial uses of its innovative education gadgets are also made by the Experimental Rural Technical School established by the Centre.

3. Publications

(i) Environment Science Series

This is a well-researched and colourfully illustrated series of books published by the Centre for Mass Education in Science with financial assistance from UNICEF. The series deals with the immediate environment of our children with materials that are related to the curriculum but can not be given full treatment in the present textbooks. So far nine books in the series have been prepared and six have been published with the assistance from UNICEF.

The books published so far, are:

Dangai, Jole Hawai Chole	(On Transport)
O, Amar Desher Mati	(On Soil)
Ay Brishti Jhepe	(On Weather)
Pani, Jake Jibon Bole Jani	(On Water)
Shabujer Shate Mitali	(On Plants)
Dhono Dhanney	(On Rice)

(ii) Other Books

(a) The Centre has authored a book on science projects for young Students, namely Khude Bigganir Project.

(b) It has edited and supervised in the publication of the following books.

Jot Pore, Pata Nore   (On small things in nature)
Ascharya ar Ascharya   (Wonders and wonders)
Biggane Boro Manush Boro Kaj   (Great men and Great works in
                                        Science)

(c) The Centre has helped in the research and publication of a unique book Bijnan Club (Science Club) published by the Bangladesh Association for Scientists and Scientific Professions. This has been a comprehensive manual for the Science Club Movement in the Country.

(d) The Centre has published a documentary booklet on Nobel prizes, in connection with its exhibition on the same subject.

(e) The Centre has compiled and edited a book on the various nonformal education and rural development groups participating in a workshop organised by Bangladesh Association for Community Education and Swanirvor Bangladesh.

(iii)   Bijnan Samoeeki, The Monthly Publication

Monthly Bijnan Samoeeki, the associate periodical of the Centre, has been regularly published for the last 16 years. Apart from its contribution to the popular science literature and the Science Club movement, the monthly has been engaged in creating useful materials for curriculum development. Actually the voluntary organisation the Centre for Mass Education in Science grew up around this periodical and drew heavily on its experience and research. The Bijnan Samoeeki, during its long efforts, has succeeded in creating a wide readership through Bangladesh, specially among the young people. A very useful feed-back from them nourishes the periodical. The periodical has two aspects one having direct appeal to the children, while the other is addressed to the more serious students, educators, and the public in general. But the style is a popular and attractive one all the way.

This periodical is the backbone of its associate science club chain ``Anushandhani'' started in 1973. The club has over fifty branches throughout Bangladesh, many of them in rural areas. These are active clubs of mainly teenagers, taking up projects exploring, surveying, building and improvising various things related to science and life.

4.   Experimental Rural Technical School

The Centre for Mass Education in Science has set up an experimental rural technical school in the village Suruj of district Tangail. The school which has been started with the help of the local people and with financial assistance from Terre des Hommes—France, is for the most disadvantaged children of the village. The site was selected in Tangail, so that the school can focus on the same landless groups being served by the Grameen Bank Prokalpa, the rural credit programme of Bangladesh Bank. The children are those, who either could not go to a primary school or had to leave it, because their parents could not afford that education. Some of the same parents however, manage to send their boys and girls to this experimental school, because it is trying to give them a package of life-oriented education as well as a skill training, that may prepare them as an income-earning member of the family in a few years. In any case, the school is right at their door-step and it is quite consistent with the village life in its form and content.

A new life-oriented condensed education package has been devised for this purpose, and it is being perfected now. The package includes:

(i)   Literacy and the use of it.
(ii)   Basic Calculations and drawing.
(iii)   Appropriate Science and Technology with emphasis on health and agriculture.
(iv)   Knowledge about Society.

Only a part of the teaching takes place in the classroom situation, the rest being taken care of through practical and project-works in and around the school. The syllabus as well as the project works have been designed to be relevant in the immediate life environment of the students. Students are helping in an important way in the creation of the educational materials and in the up-keep of the school through their project-works. The education and the practice of the same, go here almost hand in hand. Things such as the school vegetable garden, minipond or the workshop are attracting the students a lot.

On the skill training side, a few trades such as carpentry, metal craft, mechanic's craft and sewing have been taken up so far. The teachers responsible for both the general and technical education are themselves technically educated persons and they are helped by some local artisans on part-time basis. The policy is to keep the students fairly close to the real market situation of their trade, even when they are still in the school.

As the whole effort is an experimental one, the idea is to see how good such concepts work in practice. It is hoped that such an education will improve the quality of life at home and give the students an educated start in life within a short time, without alienating them from the family and the rural background in doing so.

5.   Television Series on Mass Literacy

The Centre for Mass Education in Science has devised a new TV Series "Akkar Chakkar" (The Matter with Letter) as a weekly magazine programme on Mass Literacy. This is a completely new conception in our TV as it uses many separate situations each of only a few second's duration. The content, the shortness of duration, the rapid changes in situations and the element of surprise try to make the series interesting to all sections of viewers while retaining its education value to the target group, namely the illiterates.

A single day's programme concentrates on a single letter of the alphabet. This is familiarised by means of association of ideas with interesting forms and situations reminding of the letter. Also presented are some of the simple words that can be formed with the letter. Familiar scenes, jingles, rhymes, parodies, short dramatisations and a contrasting presentation of personalities, try to jolt the viewers into an enjoyable attentiveness to the main focus, that is the letter-form of the day. A minor aim of the programme is to emphasise on some information and values in a not too conspicuous way.

The Centre for Mass Education in Science has researched and created the programme, that has been produced by the education TV section of BTV beginning from November 1981, as the weekly magazine programme "Shikya Bichitra".

6.   Some other Activities

(i)   The centre started a feature service to write and contribute features to national diaries and weeklies. The features are on subjects of science and technology related to our everyday life and development process.

(ii)   The centre organised an exhibition on Nobel Prizes in collaboration with the Nobel Foundation, Sweden, and the Swedish Embassy in Bangladesh.

(iii) The centre offered its expertise in editing and supervising some of the popular books on science and technology published by a commercial publisher.

7. The Management

The Centre for Mass Education in Science is managed by a Board of Directors. The Executive Director is the Chief Executive.

The problems inherent in the transfer of technological information in the Third World countries, and the increase in the dependence of such countries, have been described by Ruhi Sharif of Jordan:

Problems of Technology Transfer

The usual channels of transfer started the process of transfer without regard to local conditions, capabilities, needs and future impact. Everyone looked at the process from his own point of view, but all were encouraged or justified by the fact that developing countries need something additional to the traditional. Developing countries need houses, roads, food, rails, etc. and traditional technics and local materials are no more efficient and/or sufficient.

Therefore, transfer is necessary and they need to take from developed countries. But the transfer was blind, and it took no consideration of local conditions, local resources, or local capabilities. The result was floods of new technologies which drained the wealth of the country. It is true that imported technology solved some problems of the society but, instead, many other problems were created. Capital and foreign exchange was drained, investments were not responding to the needs of the society. It did not train people and it did not promote the process of socio-economic development. Investors, for instance, invested in one sector, and refused to invest in other sectors, according to the profit. Improper allocation of resources became prevailing.

The ultimate result was more dependence on developed countries and this created more burden on developing countries. As a result, complete factories were imported and their raw materials needed for production were also imported. This meant continuous flow of capital. Some projects were given as turnkey jobs to well-developed contracting companies from abroad, the company finished the work without leaving any effect such as training of local staff. It did not train people. When similar projects were needed, no local company could do the job and they have to ask the help of foreign companies.

The transfer did not create growth, and it did not promote development.

Appropriate Technology Transfer

In such a situation, there should be a new approach to the whole process. Transfer should be controlled and organised. It should have objectives and plans through which objectives could be achieved. By this, technology becomes useful to community and drawbacks could be avoided.

The World Federation for Engineering Organisations (WFEO) realised this need and it established a committee called the Transfer of Appropriate Technology Committee (TAT). This committee is one of the permanent committees of WFEO. Other permanent committees are information, Training, Energy, Environment and Promotion of Engineering Association Committees.

TAT committee comprises members from all over the world representing Engineering Associations. It is presided over by Dr Ruhi Sharif of Jordan.

This committee has been formed to organise and mobilise the process of transfer of technology among national members. Developed countries could give know-how and information to developing countries. Developing countries also could give information which could be useful for developed countries. The main idea is to promote exchange of know-how, information, technology, experience, pilot projects etc among all countries for the benefit of the whole. The committee has set out the following objectives for the process of transfer:

1. It should satisfy the needs of the community.
2. It should make maximum use of local materials, local resources and local skills.
3. It should generate economic growth and socio-economic development.
4. Local skills should be able to do the work.
5. It should reduce import and increase export.
6. It should suit the local conditions, local traditions and local environment. It is not copying, but studying, assimilating and adapting.
7. It should be easily applied, manufactured, used, exported and developed.
8. Local factories should be established for the application of technology.
9. Transfer should be accompanied by education, training, research and applications in transfer.
10. Technology should be educated to people, who should accept it and work to adapt it.
11. It should reduce dependency on developed countries.
12. It should reserve the foreign exchange.

How to Achieve These Objectives

The Committee sees that objectives could be achieved through:

1. Science and technology education at all levels, university, community colleges, schools, training centres etc.
2. Exchange of information on seminars, conferences, inventions, pilot plants, experiments etc.
3. Inform and train the public; people should understand technology, work with it, produce it, and be prepared for it.
4. Vocational training.
5. Research and development.
6. Pilot projects.
7. Continuous meetings of committee members.
8. Seminars, conferences workshops, lectures.
9. Continuing education.
10. Bulletins and news letter.
11. National committee for science and technology for development to set priorities and work for achievement of above-mentioned objectives.
12. Continuous communication with other permanent committees of WFEO.

Science and Technology Education

One of the most important prerequisites for the process of transfer is education. Through education and training, scientists could arrive at results and then convince the public that results of scientific approach could be applied and adopted. People should be convinced by new technology through experiments and then through application of results. If they participate in the application and production they will be convinced more.

They should feel that science and technology will aid and speed up the process of socio-economic development. Sometimes, people think that new development in the field of science and technology will create unemployment. Moreover, many graduates such as engineers think that after graduation, they do not need any more education. On the contrary, they need continuing education. They need Seminars, Conferences, workshops etc in order to become acquainted with new developments in technology.

This does not mean copying of imported technology, but it means that scientists and engineers should get acquainted with new technology, study them, assimilate them and adapt to local conditions.

Traditional Education or Developed Education

Traditional educational technology and approaches are no more adequate. Education should be developed to suit available conditions. Developing countries need new technology but this should satisfy certain requirements. Education is very important in the process of transfer. Therefore, it should be appropriate to suit the appropriate transfer of technology.

In education we need staff, equipment, aids, information, experimentation, presentation, application and training. Scientists should be highly associated with industry. There should be links between education and industry. Industry needs scientists help but scientists and people of education need to be aware of the industries needs.

We have to educate our people to be able to be of benefit to industry.

University should be linked with industry by knowing needs of industry. Industry should participate in programs and syllabus. Training of students should be on site; problems of industry should be solved through joint research projects carried out by both parties.

5

Global Learning: A Challenge

EDWARD W. PLOMAN

THE THEMES of this conference are closely related to major concerns of the United Nations University (UNU). Science and technology education in relation to future human needs we would tend to see as a crucial aspect of the dissemination of knowledge which in the University's Charter is given the same importance as research and training, and which, set in a wider context, has been expressed in the concept of global learning. The following observations will, therefore, focus on major selected aspects of the concept and practice of global learning in the perspective of dissemination of scientific and technological knowledge in response to human needs, both current and future.

The Concept of Global Learning

First, some indication should be given of what the concept of global learning is supposed to signify and how the UN University is trying to develop and apply this idea. The concept of global learning represents an emerging confluence of diverse trends in thinking and in practice: it is like an evolving web of ideas and activities. Drawing upon the admirable ambiguity of the English language, the term global learning can be seen as referring both to:

— learning about global issues and processes

— learning processes globally perceived

The concept "global" as used in this context cannot be equated with the traditional concept "international": it has deeper and wider connotations. It requires a complementary linkage of issues: horizontally not only across disciplines and professions but also across cultures, societies and ideologies; vertically across levels, relating the global perspective to the local, national, regional and international levels. The linkage between "globality" and learning can then be seen as the learning required to cope with global issues, from disarmament to development; thus, as Rector Soedjatmoko has stated in a recent lecture: we need to understand not only the learning dimension of development but also to consider development as learning, as something we learn rather than just do.

75

This, in turn implies the need to develop the capacity for learning, at the individual level and, equally and often more importantly, at the level of groups, institutions and even societies. What do we know about how societies learn? Why is it that some ideas or phenomena seem to "catch on" without anything that resembles education or learning (e.g. the transistor radio) while others, however well-grounded, however well-promoted, seem to make no headway at all (e.g. disarmament)?

In this perspective, it seems clear that all societies—be they industrialised or poor, East or West of geographical or ideological divides—are ill-prepared to deal with a swiftly changing, increasingly complex and increasingly competitive world. It is a basic fact that, all stated intentions and all available knowledge to the contrary, we have failed to learn how to deal with the problem of poverty, most intolerably demonstrated in the current African crisis. Since there have been some 140 wars since 1945, mostly in developing regions, we still seem to admit war as a means of conflict resolution, particularly as long as our boys are not involved, and agree to make them increasingly murderous by putting some of our best scientific talent into the development of arms. "The march of folly", to use Barbara Tuchman's striking phrase, is reflected in that the best we can come up with to avoid the ultimate catastrophe is appropriately called MAD for "mutually assured destruction". What capacity have we developed to cope with the situation at the end of this century, with another two billion people crowded into a shrinking global village, already beset by violence, hunger, poverty, environmental deterioration, ungovernable mega-cities and threats to our survival not only from earth but also from space?

Learning Needs

The UN University has been entrusted with the task of improving the understanding of pressing global problems of human survival, development and welfare. Thus, one focus of its activities is on the changing realities that together make up global issues and the corresponding learning needs.

In general terms, it is obvious that we must learn how to manage global, complex systems while respecting the autonomy of the processes and elements within these systems. We need to learn how to manage interdependence not only of countries but also of issues at new and higher levels of complexity and vulnerability. In the process of interdependence we have all become more vulnerable. Changes at the international level are now interlinked with changes at the national and even sub-national level, politically, ecologically, culturally and psychologically. Changes in modes of production and consumption affect not only economies but also resource use and eco-systems. Thus, we need to learn how to understand and manage systems that are, in a global perspective, marked by profound change and thus by uncertainty, instability and unpredictability. We are facing a situation of global transformation where at the same time science and technology have brought about an extension of human power to the point where it could destroy our planetary habitat and civilisation as we know it. If this destructive capacity seems a certainty we do, however, also need to learn how to deal with situations of scientific uncertainty or even controversy where action—or non-action—might have long-term or even irreversible effects.

Such general global issues and learning needs are directly linked to the development, application and understanding of science and technology. In addition, this conference could provide the opportunity for identifying the specific learning needs related to its themes. There is, though, one dimension of the global learning concept that seems applicable to all the varied learning issues raised by these themes.

Inherent in the global learning concept is the need to learn how to generate, evaluate, select and share vast amounts of new, timely and relevant information, at all social levels, from the villager to the political decision maker. Relevant is the key word but relevant to whom? If for example, we need to learn how to disseminate scientific and technological knowledge in a manner which makes sense to the end-users, must we not first learn how to listen to the end-users? Thus, do we not need to go beyond "education" in the traditional sense as a one-way communication process to "learning" as an intrinsically two-way process of knowledge sharing?

Learning Modes

We face these new and apparently often unprecedented learning needs in a situation where all countries seem to be in the throes of a continuous educational crisis, a crisis both *of* education and *in* education.

External factors affecting both educational systems and learning processes include the growth and diversity in the demands for learning, resulting from the increase in population and the shifts in demographic composition. The differences in the age pyramids—in many, particularly in developing countries a predominance of young people, in others a greying of the population—imply not only different learning needs but also differently oriented modes and types of learning. At the same time, there has been an increase in expectations, expressed in such commonly accepted goals as eradication of illiteracy, primary education for all, democratisation of higher education. Also the population explosion is matched by what is called a knowledge explosion and the exponential growth of available information, however fragmented and unequally available. Thus, it is increasingly recognised that conventional systems of education can no longer absorb the knowledge generated and disseminate it in the usual educational time-span, or respond adequately to the demands for equitable and widespread access to relevant and timely knowledge and information, nor to the learning needs caused by the rapid changes in and outdating of knowledge.

Concurrently, the crisis in education has been expressed in vastly changed and still changing concepts about education. The critique of traditional attitudes has in an extreme form been formulated in Ivan Illich's proposal for the "de-schooling of society". More acceptable has seemed the idea of education as a life-long, permanent process as envisaged in the well-known Unesco report with the revealing title of "Learning to be". Recently, a report to the Club of Rome, somewhat exaggeratedly entitled "No limits to learning", provides the idea of innovative and integrative learning, which requires the addition of two major concepts, of immediate relevance to global learning.

The first, participatory learning is intended to create solidarity in space: the aim should be to foster participation in the learning and information sharing processes, at all social levels, and at all ages. I should like though to extend the concept of solidarity in space to encompass the need to learn how to accept, understand and profit from cultural diversity.

This is applicable as much in education and learning as in other fields. The Unesco report "Learning to be" (1972) came to two conclusions in this respect: (a) education has a far richer past than the relative uniformity of the present structures, and (b) African cultures, Asian philosophies and many other traditions are imbued with values that could become a source of inspiration not only for educational systems in the countries which have inherited them but for universal educational thought as well.

It is interesting to note that "according to the Buddhist traditions of learning in particular, and the Asian religious tradition of learning in general, the present Western system of over-emphasising the children's education as it is done today, at the expense of the education of the adults, is an undesirable extreme that should be avoided. Meanwhile, the need for educating and re-educating the adults, particularly to understand human behaviour problems and to meet the fast changing needs of contemporary complex society, appears to be entirely neglected or not given the same emphasis." (Hewage, 1985).

The other feature, anticipatory learning is seen as providing solidarity in time, through anticipation as the capacity to face new, often unprecedented situations and to create new alternatives where none existed before. The problem is though that, as generally practised, forecasting and planning is conducted in terms of surprise-free, let's say rigidly utopian, projections that do not and cannot correspond to reality. Even Western science is slowly coming to realise that the deterministic, realistic, linear world image that underlies such projections is not valid for the explanation of the nature and behaviour of complex systems, natural and social. Rather we must learn to cope with undeterminate, complex processes, with fluctuation and instability as relevant and necessary properties of life, with the creation of order out of disorder, and the confusion and noise built into complex processes.

Confusion and noise have become very obvious in the larger information environment within which education and learning take place. All activities and institutions concerned with the generation, storage, processing and dissemination of knowledge and information are embedded in the complex, rapidly changing and barely understood social context that variously has been called the communications revolution or the advent of an information-oriented society. The forms and modes available for the production, presentation and distribution of knowledge and information are multiplying and affect not only the patterns of information use and learning but also the nature of knowledge. The new technologies influence the conduct of scientific inquiry: a good example mentioned by biologists is the difference in studying living systems such as cells using static photographs or moving pictures. The computer has already changed the methods for the collection, storage, analysis and presentation of data. Scientific and other information is analysed and presented in new visual and graphic modes: we introduce new time-scales through slow-motion and ultra-rapid as well as through carbon-dating; we change the spatial scale by picturing the infinitely small and the outer reaches of cosmic space. Thus, through the new information and communications technologies we are changing, at the same time, our knowledge of the world and our ways of knowing and learning.

Learning Practice

The last aspect I want to mention concerns the "how to", the *practice of global learning* on the basis of selected pilot activities which the UNU is undertaking.

A first point is that global learning as conceived by the UN University relates to but also goes beyond traditionally defined formal education. Global learning by its very nature must involve all levels of society and all age-groups, if for nothing else that it is that aggregate of individual decision making which conditions the success of or failure of say population policies or environmental preservation. There is, therefore, a need to consider and use all available means and methods, traditional and modern, and invent new ones.

The first example of the kind of activities the University carries out within the general framework of its evolving approach to global learning, concerns the most urgent and neglected level. There is much valuable information available that could assist the most disfavoured groups in developing societies, in their struggle for survival and betterment of life, those who live around the poverty line, with low levels of formal education, with little access to the information they require. How can scientific and technical information be transmitted to those disfavoured groups in a relevant and timely fashion and in a form that makes sense to them? We have started in India a project the results of which look deceptively simple but which requires a sophisticated process of translation and transformation. On the basis of up-to-date scientific and other knowledge and also on the basis of priority needs defined by local groups, extension workers, non-governmental organisations of various kinds, subject matters in such areas as health, nutrition, food preservation, house repairs, clean water are explained in "manuals" comprising simple, clear drawings and succinct accompanying texts. These manuals are then adapted into local languages—say Hindi, Tamil, Gujarati—and into local formats: wall journals, wall hangings or other traditional forms of dissemination of knowledge. Thus, through wall posters large population groups can be reached in public places where the content can be discussed without manipulation, at a very low cost. The effect of this method has been such that some local groups have adopted this methodology and are producing their own "manuals" on subject matters of their choice. A number of international organisations are increasingly interested in this technique, from Unesco and the World Bank to the International Council of Scientific Unions and this conference where a demonstration of this methodology has been arranged.

While this activity represents a modern use of traditional communication techniques, another project is based on the opposite approach: the use of modern technology for enhancement of traditional, mainly oral, forms of expression. In this case, lightweight video is used for development purposes at the village and community level, by teaching villagers and community workers the use of the equipment so as to give them a new means for expressing themselves, so that they can put forth their own views, priorities, concerns and grievances. This method also allows them to communicate with other villagers, or with the authorities.

We have also tried to extend the exchange of such information and experience of the international level. Thus, videotapes made with and by Chinese villagers demonstrating integrated rural energy system based on biogas have been shown to farmers in Guyana and following the interest expressed by these farmers, led to a project introducing biogas based energy-systems in that country.

Obviously, there are many other modes and methods of relevance in this context. Distance learning systems is one mode which is attracting increased attention in all parts of the world, both North and South. More and innovative modes need to be invented.

The themes of this conference demands new approaches to learning and education, to the identification, expression and sharing of knowledge and experience in which the potentials of communications, information, traditional and new learning methods and the arts are recognised as distinct but interrelated and interdependent facets.

Reference

Hewage, L. G. (1985) Global Learning East–West Perspectives with a Future Orientation, unpublished.

6
Transfer using Video Techniques

CHARLES TAYLOR

NOT VERY long ago, the production of video recordings for use in the classroom was difficult, time consuming and involved very expensive apparatus. Film using home-movie cameras was a little more available, but was not easy to edit and had to be sent away for processing. But in the last year or two, self-contained video cameras that incorporate a recorder for sound and vision have appeared on the market. Though they are still relatively expensive ($1200–1500), the possibilities that they open up are exciting. The Ferguson Videostar C is a typical example (see Fig. 6.1). It uses the new VHS-C format; the tape is identical with that used in standard VHS recorders but is presented in the form of a compact cassette which runs for 30 minutes (see Fig. 6.2). The cassette fits straight into the camera and the whole unit weighs about 2 kg. An

FIG. 6.1. The Videostar camera-recorder.

ingenious adapter is available which permits the compact tape to be played
back in a normal VHS player.

FIG. 6.2. The compact cassette.

The camera has a zoom lens (8–48 mm), a macro-focusing system down
to about 1 cm from the lens, instant playback through the black-and-white
electronic viewfinder (with sound through an earpiece) or the camera itself can
be used to drive an ordinary colour TV via an RF converter. Filters are provided
to adapt to daylight or to artificial light, automatic colour balance and many
other features are included. The camera is really intended for the domestic
market, but, it seems to me that it will have a great future in the classroom. The
price should at least put it within reach of centralised resource centres.

The production of video material raises some interesting problems. Who
should make the programmes? If we are talking about a short, single concept
sequence then a teacher with a little skill can probably do it for himself. The
great advantage then is that no one knows better than the teacher what points
are to be made, and the possible change of emphasis, or even distortion, that

can so easily be introduced by a "middle man" are avoided. (The aperture mentioned in the third section of Chapter 1.)

But, if a longer programme is to be produced, then the question of bringing in the TV experts arises. Many of us have had experience of working with professional producers and, unless one is exceptionally fortunate, one finds that the aesthetic, dramatic or technical consideration can all too easily become paramount and push the scientific teaching points into second place.

The argument is often put forward that both children and adults—especially in the developed world, and increasingly in developing countries—are used to seeing polished productions on the public television channels and are therefore less ready to accept anything that seems to be "amateurish".

I am not convinced that this is necessarily so; indeed I have argued that some television science programmes are too polished. The impression is all too easily conveyed that all experiments work the first time and that there are no apparatus problems, and then, when the students try things for themselves, they easily become discouraged if they cannot get the experiment to work instantly. From the learning point of view it is often more useful if the experiment does *not* work immediately, provided that the teacher is on hand to encourage questions about why this happened and how to correct the position.

In the audio-visual age of the 60s in the UK there were many arguments about whether the need was for AV specialists to help the teacher, or to take over from the teacher in the production of material and this continued into the period when the title changed to "Educational Technology". The argument goes on into the video age and there is still no simple answer. Clearly teachers must be involved, but also the complexities of the hardware demand the presence of the expert. Perhaps, to steal a term from the world of computers we need more "user-friendly" hardware and even "user-friendly" experts.

Apart from replacing film as illustrative material in the classroom how else can video material be used? Steve Landfried of Stoughton High School in Wisconsin, USA has introduced the idea of students using video material to present the results of projects in place of reports or term papers. Students work in pairs with help from the supervisor and produce a ¾ hour programme on a topic of their own choosing. Typical topics have been, "School budget cuts: Cost v. Quality", "Cambodians in Stoughton", "Stoughton: Toxic blues". The results have generated considerable public interest and so not only serve to transfer all kinds of information to the students, but also to the community. Steve Landfried writes

Public awareness of the student video term papers was stimulated by our local newspaper, *The Courier-Hub*, through a series of articles and reports, replete with photographs. Each documentary then premiered before a live audience for comment and discussion—a kind of public-issues forum. The forums were video-taped as well by Stoughton Community Television and were later telecast on the local community access channel. Follow-up articles and an editorial appeared in the next issue of *The Courier-Hub*.

Local coverage, however, was only the beginning. Both WISC-TV and the state public radio network featured the programme on assimilating 64 Cambodians into the Stoughton community. Madison's evening paper, *The Capital Times,* gave the toxic waste programme and forum front-page coverage. In early April, the Associated Press nationally distributed a feature article about the entire project. The National Education Association anticipates satellite distribution of portions of each programme and forum to affiliates in 21 states. And participating students may soon give testimony to state and/or federal legislators about the implication of their investigations for bills dealing with property tax relief, toxic waste management, and refugee aid.

State and national interest has focused on three aspects of the project. First, people seem intrigued by the creative outlet that the visual medium offers students as they do community-based research on vital local issues. Second, others, particularly educators, are excited that academic skills traditionally associated with written term papers can be developed and expanded upon in a form easily shared with policy makers and the public.

Finally, the project has triggered peoples' imaginations as they seek ways for schools to publicise their efforts in the local media. In-depth student research of complex community issues and public discussion of the findings offer both print and non-print media an opportunity to highlight school efforts. In Stoughton, students found that thoughtful, well-balanced dialogue about controversial issues has served the school, community, and society well.

Some of the larger video-recorders can be adapted to take time lapse sequences (i.e. single frames taken at intervals of several seconds, played back at normal speed so that, for example, a day's observations could be compressed into a few minutes). There are many possible applications, but one particularly imaginative one has been carried out by Graham Sumner, in Wales. He controlled a U-matic video recorder attached to a video camera by means of a BBC microcomputer to take time lapse sequences of cloud patterns. The computer was also linked to various sensors measuring temperature, humidity, wind speed and direction, etc., so that the results could be displayed alongside the cloud pictures. Thus it was possible to see in a short time how the changes in the various parameters influenced the cloud patterns. The following is an extract from his paper in Volume 8 of the *Journal of Geography in Higher Education:*

There is clearly a future in the use of such systems for teaching and research purposes. Microcomputer based systems have effectively already arrived on the scene; all that remains is for us to catch up and use them! Video technology is slightly further behind, at least in realising its educational potential. Video discs will certainly present totally new horizons and offer a dramatic improvement in image quality, although recorders are at present very expensive and unlike tapes, discs cannot be re-used. The problem as always is one of money, but it is also one of awaiting the general arrival of the appropriate technology in the right form and at the right price. What is technologically possible is the marrying of video and micro displays on to the screen, using split-screen techniques, and synchronising either the information gathering or its display between video and micro. Both types of display can of course be pre-prepared and recorded on to conventional video tape, as teaching packages if necessary, although this tends to produce the moving equivalent of the textbook, and leads us away from the greatest teaching potential—the "immediacy".

Videos and micros together, in all their various uses, offer tremendous opportunities for improvement in the teaching of geography particularly in those parts in which other techniques are ineffective. As well as the application in meteorology and climatology illustrated here, certain slow fluvial and tidal processes, biogeographical laboratory studies, flume studies, traffic and pedestrian flows and so on are also suitable cases for treatment. But these are left for others to write about.

Various other attempts have been made to link video players with computers, for example in interactive computer-assisted learning systems. It would obviously be helpful if, when a student is following an interactive programme, a short sequence of video material could be inserted to amplify a particular point. The main problem is that it takes a considerable time to run through a tape to find the relevant items and, of course, in an interactive arrangement, the order in which different sequences are required will depend on the student; some means of very rapid selection of the correct point on the tape is therefore needed. Video-disc seems to offer much more exciting possibilities since the amount of material stored on a disc and the speed with which it can be located are both far greater than for tape. The following extract from a paper by Martin Brown from Ireland describes a system with its advantages and limitations.

One component of information technology which is beginning to make a considerable impact on the learning of physics at 14–16 is the combination of TV screen and either videotape player or microcomputer. Many videotapes are available from the BBC, ITV companies and commercial organisations which, if used appropriately, can enhance the teacher's explanation of concepts. However, this is passive rather than active learning: almost the only option available to the student or teacher is to rewind and show a series of frames again, either at normal speed or in slow motion.

The use of the microcomputer keyboard allows the student to interact with material on the screen. Programs are being developed which can guide the student through learning material in the way best suited to his learning style and speed. He can vary parameters within a problem: effectively to ask and answer the question "what if . . .?" He can look up tabulated information such as densities or melting points or today's prices of materials. He can work through case studies illustrating commercial applications of the concepts he is learning. And he can use the microcomputer to monitor readings in an experiment, or to control the operation of the apparatus. However, although some initiative has passed to the student, most programs are still closed-ended in that in general the programmer has ultimate control of the information to be used in the program.

The versatility of the system can be increased by bringing the videotape player and microcomputer together, and several organisations are preparing microcomputer-controlled videotapes which allow the student to learn by holding a conversation with the tape through keyboard and screen. The technology is limited by the long time interval (up to 5 minutes) required for the microcomputer to find appropriate sections of tape, and by damage to the tape when single frames are held on the screen for more than a few seconds.

These problems can be avoided if the learning material is stored on videodisc rather than tape. Information is permanently recorded on the videodisc in the form of millions of small closely-spaced pits in its surface, and is read from the disc either by reflecting laser light from its surface or using the variation in capacitance as a probe travels just above the surface. Any one of over 50 000 frames can be found within 5 seconds and, as there is no physical contact between disc and reading mechanism, there is no deterioration in picture quality with use.

Several lines of development are being followed at present for both domestic and commercial markets. Low-priced videodisc players are being produced for home entertainment. Discs have the advantage over tapes of lower cost, higher picture quality and greater durability. But they have the great disadvantage for home use that, unlike tape, it is not possible to record directly on to disc. The disc must be bought as a finished product. The rate at which interactive videodisc can be introduced as the basis of home education may therefore depend on the rate at which videodisc becomes established in home entertainment, as the development costs of specially-designed discs are high and there are not yet enough videodisc players in homes for potential publishers to consider such investment worthwhile. In commercial use the interactive videodisc is being developed mainly for personnel training, storage of reference material, and point-of-sale advertising.

A student of average intelligence, who is sufficiently motivated, can learn the basic concepts of physics directly from a good textbook, particularly if a reasonable number of well-designed questions is included to reinforce his experience and the information in the book. Such a book may contain 200 000 words of type or about 1 megabyte (1 Mb) of information. Purely as an information-storage device the potential capacity of a videodisc is over 1000 Mb. However, in practical use not all of this capacity will be used. If the disc is to be developed mainly to store printed matter then, assuming that for comfortable reading from the screen the useful capacity of a single videodisc frame is about 80 words or 400 bytes, a videodisc of 54 000 frames can carry over 20 Mb, about twenty times the capacity of a book. Therefore, at its most primitive level, an interactive videodisc could simply be an electronic textbook, but with the opportunity to include many more questions, many more illustrations in colour, and replace many still diagrams by short moving sequences and animation.

The basic disc would contain words on screen simply read in sequence by the student, with still or animated diagrams at appropriate points. On completion of each section there would be a test and, as a result of this, the student would either go on to the next chapter, or repeat selected material at a simpler level before taking the test again. An accelerated level would also be available so that, on revision, the student could be simply reminded of the material in outline before taking each test.

However, simply to reproduce slightly enhanced book material on a screen is not the most effective use of the technology. If combination of a microcomputer and a videodisc player can form the basis of a learning system for use in a classroom by groups or individual students. Students have access to the keyboard and screen, and also have available experimental apparatus, textbooks, writing material, and a printer to record relevant material from the screen. The microcomputer can also be linked to a national database like Prestel, with the eventual capacity to get information from international databases.

Immediately it is clear that control of the learning process has passed from teacher to learner. The pace of conventional class teaching is determined by the teacher, taking into consideration such constraints as examination demands and the range of ability of his students. To accept the disc-based system is to accept the concept of independent learning, in which the pace of learning of each student is largely determined by his own needs.

The teacher will then see his role change from director in front of the class to manager in the background. Students will be at different points in a course, though their progress can be monitored by the computer. And, as the effectiveness of the system depends on the degree of motivation of the students using it, encouraging such motivation will probably become the teacher's primary activity.

In a typical session at the keyboard, the student switches on the system with the appropriate course disc in the active player, and a disc with material relevant to the part of the course he is studying in the passive player. The active disc is designed to contain the management structure of the course; it decides, on the basis of student response, the level and speed of presentation of material, and stores a large amount of still and animated material on single frames. The passive discs contain passages of direct video material of varying length covering the subject matter of the course. Information from the first few frames of the active disc is directed to the microcomputer rather than the screen: this is the controlling program. The screen first asks for a "bookmarker" code number which the student has written down at the end of the previous session so that it can start where he left off, and then presents him with a short introduction in words, moving pictures and diagrams to the next section of the course. He may be asked to decide the rate of instruction he requires and the grade to which he aspires. The microcomputer takes him through each step of the topic at the appropriate speed, frequently asking questions, and responding to his answers with relevant still or animated diagrams, short moving sequences or further questions. Longer moving sequences can be taken from the passive disc. Some of the questions will require problem-solving on paper, or experimental work and the microcomputer can monitor directly, for example, the rise in temperature of a metal block, relating this to the current and voltage in a heater circuit and the mass of a block measured on an electronic balance, to calculate and help explain the meaning of specifc heat capacity. The student is asked occasionally if he wants, for example, further information on the relevance of heat capacity to central heating, and he can be provided with such information from a passive disc. He can be connected to the Prestel database for local, national or international information. Telesoftware, in which a relevant computer program which might enable him to calculate the annual cost of heating his home from different energy sources, can be sent down the telephone line to his microcomputer. Or he could be linked directly to databases at the National Coal Board or British Petroleum for up-to-date information. The interaction continues until the microcomputer is satisfied that he has reached the predetermined level of competence, and has acquired all the additional information and experience he needs. The program can be modified so that it may be used by a group of students rather than an individual.

If the process of achieving specified learning objectives in the disc-based system is more effective than passively listening to a teacher, then in a typical week the student may spend a relatively short time at the keyboard. His activities during the rest of his time may be subject to negotiation between himself and his teacher, but they would probably relate to many of the educational aims which are presently stated in syllabuses but are seldom assessed—and are therefore ignored by teachers. These include the stimulation of an interest and enjoyment in physics, and the development of an awareness of the social, economic and technological implications of physics. If the ease of learning the content of physics results in a shift of emphasis so that there is a redistribution of time between content and other aspects of the subject, then the assessment system must be redesigned to reflect more satisfactorily the new weighting of the aims of the course. To be realistic, redesign of the assessment system may have to precede redesign of courses.

An argument has been made for the computer to take part in the assessment process, but the knowledge that he was being examined as well as being taught by the system is likely to have an inhibiting effect on the student. He must be free to make mistakes before he can learn from them.

Another of the developments of high-technology is the use of satellites to relay television programmes to whole continents. We heard some encouraging details of India's own satellite system (INSAT); the satellite itself is, of course, a multi-use device and incorporates communications circuits, meteorological apparatus and many other facilities in addition to the television relay system. One interesting development is that meteorological information can be received by a small local station where a cheap microcomputer can select out the information that is relevant to farmers, fishermen, local administrators, etc., and this could obviously be of great importance for warnings of abnormal weather conditions. The organisation and co-operation between the various bodies involved in the day-to-day use of INSAT is very impressive, and there are certainly lessons to be learned from this in other regions.

S. S. Swani, of the UGC Mass-Communications Unit in New Delhi told us of the technical facilities that are being established in order to utilise INSAT for Higher Education programmes. He writes:

The University Grants Commission as an agency concerned with all the universities and colleges and particularly responsible for maintenance of quality and standards has taken the initiative to utilise the one hour transmission time assigned to higher education in the INSAT-IB Satellite Programmes. The Commission constituted a Working Group to advise it on various matters connected with the setting up of Centres of Mass Communication and Educational Technology in Indian Universities. On the recommendations of the Working Group a Task Force was appointed in August, 1982 to prepare a plan of Action. The recommendations of the Task Force and the Working Group were considered and accepted by the Commission at its meeting held on 23 July, 1983.

The Commission is supporting the following four Educational Media Research Centres (EMRCs) and two Audio Visual Research Centres (AVRCs) for training and production of software.

Educational Media Research Centres
   (i)   Jamia Millia Islamia, New Delhi.
  (ii) Gujarat University, Ahmedabad.
 (iii) Poona University, Pune.
 (iv) Central Institute of English and Foreign Languages (CIEFL), Hyderabad.

Audio-Visual Research Centres
   (i)   Osmania University, Hyderabad.
  (ii) Roorkee University, Roorkee.

A few more Centres are proposed to be set up during the 7th Plan.

Based on the recommendations of the Task Force and the Technical Advisory Committee for the EMRCs, the minimum facilities, i.e. equipment and staff needed for programmes production were approved. The Centres at Jamia, New Delhi University of Poona, Pune, Gujarat University, Ahmedabad and CIEFL, Hyderabad have already become operational. Other centres are in different stages of being set up.

Programmes produced at these Centres are presently being supplemented with good quality educational programmes available within India or abroad for 1 hour daily telecast through INSAT-IB and Doordarshan network. Colleges and universities all over India within the coverage range of Doordarshan network have been identified for the provision of Colour TV Receivers to enable viewing of these programmes by students.

Technical facilities at the Media Centres are relatively on a smaller scale with the use of lower gauge video equipment compared to the facilities available at a professional broadcast TV studio set up. Medium grade three tube (⅔ inch) colour cameras and ¾ inch U-Matic (Low Band) video cassette recorders together with the same level of monitoring, switching, playback and editing equipment have been provided at these centres. This equipment and the format of video recording are compatible with the format used by Doordarshan for news and field coverages (ENG) and for programmes production at its INSAT Centres. Facilities at EMRCs and AVRCs are described below.

Educational Media Research Centres

Facilities at an EMRC comprise of a TV studio of about 60 square metres in size with ancillary areas for production control room, telecine, video recording, editing, film preview, set construction, etc. Carpet area of 600 square metres, i.e. square metres of technical and 240 square metres of offices has been envisaged for such Centres. However, presently these Centres are temporarily housed in a much lesser area by utilising available building in the university and modifying the same to convert into an improvised TV studio set up.

Acoustically-treated TV studio has been centrally air-conditioned. Other areas are conditioned with window type units. The studio is equipped with 2 Nos. of three-tube colour cameras. There is a provision to deploy third camera out of 2 Nos. of similar cameras (portable version) provided for field coverage together with portable video cassette recorders. Associated video equipment such as Video Switcher with special effects, teletype writer (character generator) for captions, sync. pulse generator, colour picture monitors, wave form monitors and vector scope have also been provided. 2 Nos. of ¾ inch U-Matic Video cassette recorders are provided for recording and playback. Digital time base corrector is used for playback. There are additional 2 Nos. of similar VCRs with edit control unit and colour picture monitors for editing work. A telecine island comprising of 2 Nos. of 35 mm slide projectors, one 16 mm film projector, optical multiplexer and 3-tube colour camera together with monitoring accessories are available for film inserts or for transfer of film programmes to video. Audio equipment comprising of 8-channel mixer, microphones, audio tape recorder, turntable, monitoring and distribution amplifiers have been provided. Major items of video and audio equipment have been shown in the facility diagram.

Equipment for power supply distribution, studio lighting, test and measurements, fire fighting, etc. has been provided. Facilities for still photography, film previewing, audio recording and audio editing have also been provided.

Production, technical and administrative staff has been provided to cater to limited training requirements and production of about 1 hour of educational software per week.

Audio-Visual Research Centre

AVRC in a university is provided with library facilities in educational software both for audio and video. Listening and viewing facilities have been provided.

Two sets of portable 3-tube colour cameras and ¾ inch U-Matic Video Cassette Recorders together with editing video cassette recorders and monitoring facilities are provided. This equipment is meant for training in the electronic media and for limited production of educational programmes.

An improvised TV studio is being set up in a suitable building provided by the university to enable recording of programmes with the portable equipment.

National Centre

The University Grants Commission has set up a mass communication unit in the UGC Office for implementation of the programme of development of mass communication system and programmes in the universities and to co-ordinate the activities of the Educational Media Research Centres and Audio-Visual Research Centres. This mass communication unit is co-ordinating the functioning of the Educational Media Research Centres and Audio-Visual Research Centres and looking into the administrative aspects of funding the media centres, software as well as technical aspects and the related matters.

A UGC INSAT Project Unit has been set up at the mass communication research centre, Jamia Millia Islamia, New Delhi for collecting, previewing, selecting of programmes and putting them into 1 hour capsule with a proper mix of programmes to be passed on to Doordarshan for uplinking with INSAT–IB for daily telecast. The project team is also receiving the feedback on UGC TV Programmes and analysing it.

The Mass Communication Unit in University Grants Commission and the UGC INSAT Project Team (located at MCRC, Jamia Millia Islamia) are together discharging the functions that are to be carried out by a National Centre as envisaged in the report of the Task Force. The Centre is proposed to be set up in Delhi.

The National Centre for Educational Media Research and Development in Universities (NCEMRDU) will co-ordinate and facilitate the functioning of the EMRCs and AVRCs.

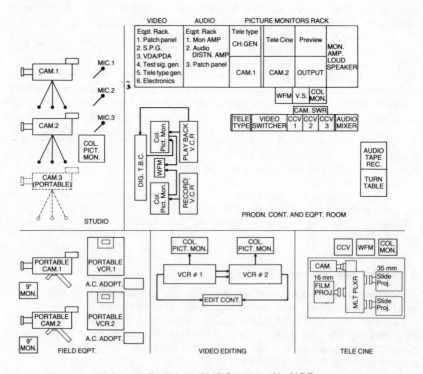

Fig. 6.3. Facilities at EMRC supported by UGC.

It will act as Information Centre and assist in procurement, duplication and distribution of audio-visual materials. It will promote innovation and research in educational technology and cater to the needs of training of teachers in the use of electronic media. The Centre will also have production facilities both for video and radio programmes and undertake production in specific areas on experimental basis.

Adequate technical facilities to carry out the above mentioned functions are proposed to be provided in the National Centre. These will be in the form of TV and sound studios, video and audio production controls, telecine video and audio recording, preview and projection facilities, editing and capsuling, equipment for duplication and distribution, TV Standards convertors, documentation and library facilities, etc.

The actual content of educational television programmes is still a matter of considerable controversy. It is tempting to see television programmes and video tapes as substitutes for teachers, or, at least, as a means of enabling a very good teacher to reach a very much larger audience than would normally be possible. But many workshop members agreed that, while this may be helpful as an interim way of dealing with a teacher shortage, TV programmes dealing with curriculum material are not exploiting the medium to the full. The real value is in linking the curriculum with real life in the community and in producing material that will enliven and alert students to the implications of what they are studying.

TV has, of course, been used in transferring information to the general public in, for example, village communities. But very often the programmes made for this purpose are of the documentary type made by outsiders who move in for a short time. A much more effective way of operating is to involve the members of the community themselves not only in the actual recording of the programme, but more importantly in the selection of topics and in the style of treatment. The same is true of programmes designed for students. Although there was by no means complete unanimity on this point the feeling began to emerge that some degree of professionalism (in the television producer's sense) could often be sacrificed in order to achieve immediacy and relevance in the resulting programme.

7

Transfer with Computer Technology *out of date*

JENNY PREECE AND ANN JONES

Introduction

At the conference most emphasis was placed on the use of computer simulations in teaching science and a double workshop session was allocated for this topic. In the first part of the workshop we discussed the reasons why computer simulations can be valuable for teaching science at both the school and university level. The main aim of this session was, however, to provide all delegates with first-hand experience of running and evaluating a simulation. In the second session Dr Claude Janvier gave a talk about designing computer simulations and this formed the basis for a discussion on what kinds of features are important in designing simulation programs. Before discussing these workshops we shall, however, describe some of the other ways in which computers can be used as teaching aids, as this generated a lot of debate throughout the conference. Much of the discussion was stimulated by the experiences of delegates in their free time. This was made possible by Sinclair Research Machines Ltd who made available several Spectrums, Spectrum Pluses and QL machines and also provided a wide variety of software to run on these machines. We were also lent five BBC Model B machines by the Indian Institute of Science and an Apple II machine by the Apple Distributors in New Delhi.

Different ways of using Microcomputers

There are many different ways of using microcomputers to assist in the teaching of science or to alleviate some of the more onerous tasks concerned with collecting and analysing data or controlling experiments. Some of these are as follows.

Statistics. For many years now scientists have benefited from having computers to do statistical analyses of their results and other kinds of calculations. Although pocket calculators obviously provide the necessary computational power to do some statistics they generally cannot be programmed to do automatically the various different kinds of computation necessary for most analyses. They are, therefore, more tedious to use than a prewritten computer program and also provide far greater scope for error.

Information retrieval. Another well-known and important use of computers is the facility to access specific items of data from large data bases. There are many ways that these kinds of programs can be used in science education such as for examining field data. (More information about this topic is contained in Chapter 8 edited by Anne Leeming.)

Word processing. Many delegates experienced the ease with which simple documents can be produced and edited using a simple word processing system on the BBC machines or the more sophisticated Sinclair QL system.

Spread sheet programs. Spread sheet programs were also available at the conference. These programs enable the user to input a variety of different figures such as costings and to see what the effect of changing one or two of the figures would be on the whole accounting system.

Modelling. Modelling programs now exist which provide friendly environments which enable pupils to develop their model. Using these programs the pupil is protected from such problems of having to scale graphical displays.

Simulations. Computer simulations are programs which simulate real events and we shall discuss them in more detail later in this paper.

Drill and practice. These are programs which generally ask students a series of questions and check their answers against the correct answers stored in the program. If the answer is correct then a simple reward such as a smiling face may be drawn and the pupil is presented with more questions. These are the simplest programs to write and they are also amongst the most commonly available to the home computing market. As a form of education they are usually didactic and rather dreary.

Adaptive drill and practice. These are drill and practice programs which contain a model of the task and also develop a model of the student. These two models are used in conjunction with each other to select tasks which are appropriate to the student's learning needs. Sometimes the models are very simple and do little more than count the number of correct responses that the student makes in any particular session, but they can also be very complex.

Tutoring programs. These are usually large programs which require a mainframe or a microcomputer with a large memory to run. These programs were first developed in the sixties in America for use on large mainframe computers. They were written to be used by large numbers of students. The programs themselves were designed to provide tutorials and these tutorials are written in specially developed programming languages called authoring languages. Authoring languages are, in theory, written for non-programmers so that teachers who have little experience of computers can write their own tutorial programs. In practice it has been found that most authoring languages are not so simple to use and those that are fairly easy suffer from the trade-off of being inflexible. The teaching material is usually presented in frames and the dialogue leads the student towards some instructional goal by skilful questioning, prompting and guiding. One of the most well-known systems is the PLATO system which is now available on micros and which uses the TUTOR authoring system.

FIG. 7.1. Workshop session on microprocessors.

Direct data capture and analysis. Little mention was made of these uses but as we are directing our discussion to science teachers we should not forget that these are important ways of using computers in research and university science and hardware and software is now being developed to enable schools to do this kind of work. Displaying images on the screen using the high resolution graphics was considered valuable for teaching many concepts but it was discussed mainly in relation to simulation programs.

Control of apparatus. The use of computers to control other apparatus was not discussed in the workshop, but this is assuming more and more importance in science teaching.

No mention was made of the importance of teaching particular programming languages and this is perhaps not so important in this context as most science teaching in schools is likely to be based on prewritten educational software.

Having briefly reviewed some of these different types of applications we now return to our discussion of simulations and the actual events and discussion that took place in the workshops at the Bangalore Conference. We shall start by first considering why simulations may have an important role in the teaching of science.

Why are Simulations Useful?

A computer simulation is a representation of reality. The computer program contains a mathematical model which is designed to make the program behave in the same way as the real system which it represents. Often the model has many limitations as real systems are usually very complex and difficult to model or there is not sufficient known about them to construct really accurate models. This is an important teaching point and one that both teachers and students must take into account in their use of computer simulations. Having mentioned these limitations, there are, however, many reasons why computer simulations provide a valuable teaching role. Computer simulations provide students with the opportunity to investigate and experiment with systems and processes which:

—are very complex and have many interacting variables;

—happen over very long or very short periods of time, such as evolutionary processes and chemical reactions;

—are dangerous;

—are undesirable or unethical, such as pollution studies on population control experiments;

—are expensive, either because of the techniques and equipment needed or because of time or the distance to the site;

—require sophisticated experimental techniques which cannot be mastered easily by students;

—must be replicated many times in order to produce meaningful results and which would, therefore, take too much lesson time. (For example, it may be considered important for biology students to do breeding experiments with Drosoabila flies, but it takes a long time to breed several generations and there are often problems with the cultures getting contaminated or drying up.)

Computer simulations make it feasible to explore "what would happen if" questions raised by students without taking too much time out of an already tightly scheduled lesson plan.

Computer simulations are not a panacea for learning, however there can be many problems. The simulation must be well designed, easy and flexible to use by both teacher and students, the model upon which it is based must be realistic and accurate, and the simulation must be well integrated in with the students other learning activities. In order to promote discussion of these and other pertinent issues we designed a practical activity for the delegates at Bangalore to participate in and we shall now describe this activity and the discussion which followed.

The Simulation Activity

Task Information

In this simulation (RELATE), the group takes on the role of the Medical Control Officer of an African village called Agwin in Northern Nigeria, and the group's task is to reduce and control the number of people infected with Malaria at the lowest possible cost. The constraints within which the Medical Control Officer works are as follows:

(i)   a 6-year contract which can be extended for a further 6 years.

(ii)   a budget of $1200 for the 6-year period.

(iii)   up to three control measures can be used in any 1 year. (The effectiveness of the controls varies according to the different stages of the parasite's life cycle and the different times of the year. The controls also have different costs.)

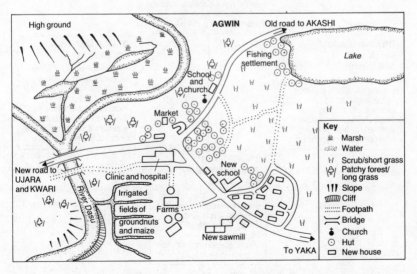

FIG. 7.2. Map of an African village called Agwin.

There has been no Medical Control Officer in Agwin for several years but 90% of the population have had regular blood samples taken, and the Medical Control Officer has access to the endemic level of the disease at the time of his

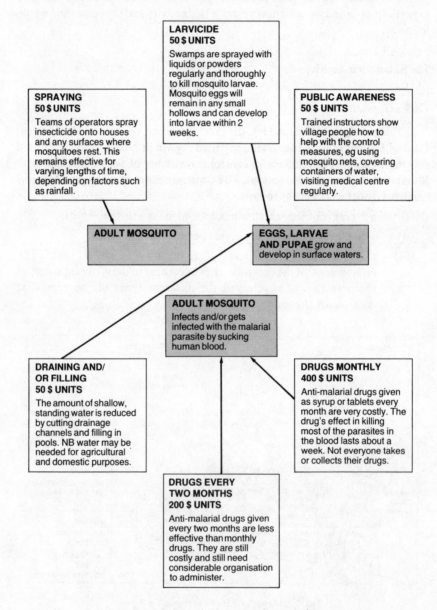

Fig. 7.3. Control measures against malaria.

or her appointment. The Medical Control Officer also has relevant information about Agwin: population, climate, fauna and geology, and a map of the village.

The Video

Like most simulations Relationships is intended to be part of a total learning package.

The life cycle of the malarial mosquito is complex hence the difficulty of controlling the spread of malaria. Before using the simulation participants watched a video, designed for use with the package, which describes and illustrates the complex life cycle of the malarial parasite, and discusses the pros and cons of various methods of malaria control.

The Environment of Agwin

As mosquitoes breed in standing water it is important to have this information for Agwin. Agwin is at a height of 500 m, and has a population of 500. It has no rain in the dry season and 440 mm in the wet season. It is situated in a late drainage area. Marshes are formed in the wet season but dry out in the dry season. Surface water collects in the wet season, and may remain after the rains in areas where clay has been extracted for building. This provides an ideal breeding ground for the mosquitoes. The enclosed map is also provided.

Malaria

Clearly, one of the most important items of information concerns the disease itself. The video has already described the disease and the life-cycle of the

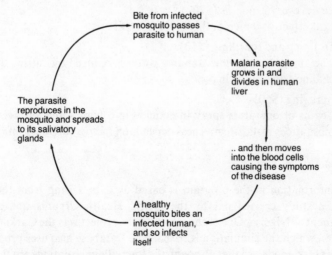

FIG. 7.4. The interaction between the parasite, mosquito and human host.

parasite, and the Relationships booklet provides groups with notes on the nature of the disease which are summarised below:

—A single cell parasite is transferred from person to person by the female mosquitoes.

—Symptoms include fever. Red blood cells are damaged. If untreated attacks recur and are very weakening.

—The interaction between the parasite, mosquito and human host is shown in Fig. 7.4.

Control Measures

The control measures operate at different stages of the material parasites life cycle—and incur varied costs. Costs must be kept under $450 per year, and the aim is to keep overall costs under $1,200 for the 6-year period.

The available control measures and their costs are summarised below:

1. Larvacide ($50)
   Swamps are sprayed regularly which kills the larvae. Eggs may remain in hollows and can develop into larvae in 2 weeks.

2. Public Awareness ($50)
   Trained instructors give guidance and advice, e.g. using nets, covering water, regular health checks.

3. Drugs monthly ($400)
   Monthly anti-malarial drugs are costly. They kill most of the parasites in the blood and the drug's effect lasts about a week. Not everyone takes or collects drugs.

4. Drugs every 2 months ($200)
   Less effective than monthly drugs but still costly.

5. Draining and for filling ($50)
   The amount of shallow standing water is reduced by cutting drainage channels and filling in pools.

6. Spraying ($50)
   Teams of operators spray insecticide into homes and surfaces where mosquitoes rest. Effectiveness depends on factors such as rainfall.

The Model

The information in the program is based on data arising from the Garki Project, a study carried out by the World Health Organisation and the Government of Nigeria. One of the results of the project was the Garki Malaria Model, on which the simulation is based. The Database also uses productions from the Garki model, so that although the four village situations are fictitious, they are representative of the range to be found in and around Agwin. The information given about the model in the documentation for the program is summarised below:

—Average population of Agwin is 500 +/− 10

—80% to 95% of this population come for blood sampling

—The normal or endemic level of infection is 50% in the dry season and 60% in the wet season

—Percentage infection levels are influenced by the "unexpected events" (if these are selected)

—Control measures can be changed annually and it is assumed that they are implemented at the relevant time of year (so for example larvicide is most effective in the wet season)

—The model takes into account the fact that the effects of draining or infilling and the public awareness campaign will have longer lasting results—the time being related to the length of time these control measures are in operation

—The model limits users to a maximum of three control measures in 1 year with a maximum annual budget of $450 units. (The term $ unit is used to represent a currency)

—The program monitors progress and gives a summary with relevant comments after 6 years and 12 years.

Using the Simulation in the Workshop

The participants were divided into five groups, each of which had a leader who was familiar with the program. As time was limited groups were asked to do six runs using just one control measure—so that they could see the effect of each control measure. After this the groups explored various combinations of the control measures in order to find the most cost effective program of control.

Discussion after the Simulation Activity

Two questions were asked in order to provide a focus for the discussion, these questions followed by the points that were raised by the audience are listed below:

A. What do you learn about malaria

1. Students realise that there are many variables that we have to consider in real-life situations.

2. Commonsense strategies such as draining wet land, actually works in the control of malaria.

3. Money has to be spent in order to control malaria.

4. Control measures will improve the situation if it is practised consistently over a period of time.

5. From playing with the program for a period of time, it is evident that there are two categories of control measures: the preventive and the actual treatments of the disease. It can be seen that using preventive measures alone for a period of 6 years actually pulls down the infection percentage.

B. Advantages and disadvantages of using the simulation in teaching

1. Participation in a simulation game is a good way of sensitising children to the problems people involved in health measures have to consider.

2. This simulation technique can be used to teach about other diseases besides malaria.

3. Whether this simulation is successful depends on what you want to get out of it.

4. The program is not flexible enough to allow for a different strategy to be used in the wet and dry seasons.

5. Many details of the dynamics in the process of malaria control are missing: how the larvae population changes after certain actions have been implemented, how many of the population actually died each year, etc.

6. There has been very little information on how the model has been built up. This is because:

    (i) sometimes the models used may be quite questionable.

    (ii) the publishers do not want to divulge the model because of commercial considerations.

    On the other hand, it would be very helpful if we can discuss the model with the students, and alert them to the areas where the model has not been realistic.

Following on from the discussion of these points we discussed ways in which the simulation can be used to best advantage and these are:

1. The teacher should highlight which are the preventive and which are the curative measures.

2. The students should have thought very carefully and bring with them planned strategies before playing the game. This way, it would be less time-consuming and also the children can learn more.

3. It may be a good idea if the teacher can suggest different possible strategies, e.g. preventive and then curative or curative measures first and then preventive.

4. It is very important for the teacher to collect the results from different groups and discuss with the children why do certain strategies produce better results.

5. The schoolchildren are not actually the people who can implement the control measures. These software designed for schoolchildren in the first world may be better used in tertiary institutions and health training centres in the developing countries.

Design Considerations

In the second workshop session we considered some of the important features of designing computer simulations. Claude Janvier gave a presentation which provided the basis for our discussion.

Claude Janvier began his talk by considering some of the reasons why computers are valuable in education. Stonier (1985) has also summarised these realms in his paper.

1. The most important reason is that the computer is interactive—unlike books, tapes, films, radio and television, the user's response determines what happens next. This gives children a sense of control. It also elicits active, motor involvement.

2. Computers are fun. Human beings love to respond to challenges and they love to make things happen. The computer games industry has grown rich on that basic axiom. By coupling education to games of challenge, computer-assisted learning becomes fun.

3. Computers have infinite patience. A computer does not care how slowly the user responds or how often a user makes mistakes. Among the earliest uses of computer-assisted learning on a wide scale, was the use of computers in Ontario in the late 1960s designed to help innumerate teenagers to fulfil the maths requirement for college courses. The program was successful from a number of points of view, not the least of them the attitude expressed by one girl who stated that the computer was the first maths teacher who had never yelled at her.

4. Good education programs never put a child down. Instead they provide effective positive reinforcement.

5. Computers can provide privacy. Children, or for that matter teachers, can make embarrassing mistakes without anyone seeing them. Ignorance, lack of skill, slowness to comprehend, poor co-ordination, all can be overcome in the privacy of one's home. The computer won't tell.

6. At the same time, the computer can be used in a variety of social situations. These include classroom activities involving groups, or a teacher and single pupil only, or two neighbourhood children, or party games, or a grandparent and grandchild, etc. Many education programs

are designed to allow for either individual practice, or for two or more children to play games.

7. The computer can explain concepts in a more interesting and understandable manner by means of animated material. No amount of talking, writing or providing diagrams, can compare with making things come alive on the screen.

8. Whereas it is very difficult to hide things in a book, it becomes possible to hide things in a program which becomes apparent only on occasion. A book on re-reading holds few surprises (although the reader may have missed points the first time). In contrast, a computer program can be full of surprises. Good programs contain an element of mystery and uncertainty which keep the user interested. It means that the learning experience provides new situations not only for the students, but for the teacher or parent as well.

9. The ability to simulate complex situations such as chemical reactions, ecosystems, demographic or economic changes is a particularly powerful reason for using computers in education. Training pilots, managers, doctors, chemical engineers, i.e. any profession or activity where a mistake in the real world could be very costly, is best served by learning on a computer which simulates the real-life situations. In addition, simulating real events often makes it possible to train students to think "laterally" across traditional subject boundaries.

There are, however, some disadvantages as Schenk (1985) points out:

Computers need software in order to teach, and in fact there are throughout the world for any level of education. Nevertheless, scattered around the world individuals have written educational programs of high quality, but these are just used and known locally. Thus it is very difficult for individual teachers to search for, to find and to lay hands on quality programs which meet their pupils' needs.

What is lacking, is a framework which guarantees the quality of teaching software, easy access to complete catalogues, and, sometimes, affordable prices. Maybe here is a task for Unesco, because certainly the fact that computer-driven teaching is independent of teachers, implies that when a good program is publicly available for a reasonable price it can be applied in any country and so it will help to spread out knowledge equally over the world.

Claude illustrated some of the aspects that he considered important with reference to some of the simulations that he has developed in Montreal. First, however, he attempted to define more precisely what he meant by a simulation. He said:

Simulations can be viewed as a model at work. But what is a model? It is a schematisation or representational version of an object, a system or a phenomenon. It is therefore, abstract. A model can be roughly described as a dynamic representational version of an object, or system which actually illustrates the interrelationships between the parameters involved. (For more details see Djordjija Petkovski's (1985) paper entitled "Computers and the World of Large-scale systems".

Simulation programs can be classified according to two criteria:

1. *The type or level of schematisation*
   These include the following kinds of representations: pictures of reality, numerical versions in which the user communicates by specifying values, diagrammatic versions, and graphical versions.

2. *The kind of interactive features*
   (a) the opportunity to intervene during the process.
   (b) the mode of interaction, e.g. awkward uses of the keyboard, systematic uses of the keyboard, direct contact with visual display, joystick, mouse, paddle, or light-pen.

One of the problems of creating good teaching programs is that you have to take into account the kinds of primitive conceptions that children have and how they will interpret the screen display. The children's interpretation may be quite different to what you, the designer expects. It is important therefore, to thoroughly test the program at various stages during its development which can be a very time-consuming process, but it is essential if you want to produce good software. Another consideration, is how much the pupil will be involved in the learning process. Will the pupil acquire knowledge passively as is often the case when reading books or listening to a teacher talk, or will the learning process be active. Active learning requires a reconstruction of knowledge and in this process the knowledge becomes personalised. Designing effective displays is also difficult. People often think that visual displays are a panacea for learning but this is not always the case. Research has shown that motion and transformation are difficult to apprehend because:

(a) With static images conventional codes and symbols are misunderstood. Arrows, for example are nearly always interpreted as indicating direction when they may in fact be indicating the next stage in a process or the point at which a particular event takes place.

(b) With dynamic representations there may be several elements involved in the motion and it is not easy to focus on the right element at the right time and in relation to other elements.

There is also of course, the question of how to use colour and symbols so that they assist the pupil's comprehension and do not mislead her. Much more research is needed on the design of visual displays. Four aspects need to be taken into consideration:

1. the content level (and perhaps several different forms are needed to cater for different pupils' needs);

2. the teacher's intervention level;
3. the "interactivity" in program control level;
4. the colour, size and position level.

Some of these points were illustrated with reference to particular examples. One of these was a simulation in which pupils form hypotheses about the effects of different amounts of a liquid upon a "diver". They are presented with a display and they are invited to change the apparatus to test their hypotheses. The parameters which they can change include the amount of liquid, the density of the liquid, the height of the measuring cylinder and its circumference, initial pressure, etc.

Discussion of design considerations

The concluding discussion focused upon three main themes.

A. Definition of computer simulation

1. What a simulation is.
2. We must define what we want to model before we can start modelling.
3. Simulations should, if possible, show the dynamics of the processes being simulated.
4. Whether or not a computer is necessary.

B. Points one should look out for in using simulations

1. We have to warn students that models should not be mistaken for reality.
2. We should try to let students see the experiment itself at work and not let simulations replace experience of the processes studied. Moreover, we should in all cases provide the necessary theoretical basis of the simulation before students are introduced to it.
3. Models are never perfect or complete—they are only complete in their context of conceptualisation.
4. Testing of simulations—experimental testing is not complete testing.
5. We have to be aware of the danger of using computers just for their novelty.
6. Computers must be integrated into the curriculum or else they will be treated as toys.
7. The use of computers should help in the development of problem-solving skills.

There were also informal discussions throughout the conference about the various attributes of the program/user interface which make programs difficult or easy to use and many of these are summarised in a paper by Anne Leeming (1985).

Concluding Remark

We concluded our discussion with a consensus of opinion that: The advent of microcomputers and the use of simulations in science teaching has provided a new horizon in educational technology. Yet as with all teaching aids and resources, how they may be incorporated into the total teaching activities is a most important factor in bringing out the teaching points that are intended to be introduced.

References

Leeming, A. *The Interface between Computers and the Naïve User.* Appendix 1, no. 5.
Petkovski, D. *Computers and the World of Large-Scale Numbers.* Appendix 1, no. 10.
Schenk, H. *On the Use of Computers in Teaching.* Appendix 1, no. 8.
Stonier, T. *Computers and the Future of Education.* Appendix 1, no. 9.

8

Looking After Data Using Computer Technology

ANNE LEEMING

Introduction

The transfer of information frequently needs to access stored data and in turn needs to add data to store. Information can be "created" by manipulating the data previously assembled and stored. Many of these processes can be extremely difficult to carry out by manual methods because they are slow or laborious so that they are simply not done. With the arrival of computers many such operations become possible and often extremely easy to carry out and many more ways of examining and combining data become possible.

Essentially the task is to MANAGE the data belonging to a piece of work or a project. That means that a policy for looking after the data needs establishing, a suitable means of carrying out this policy needs finding and then the associated tasks need allocating and the performance of the policy needs monitoring.

The topic group looked at several techniques of storing and presenting data, some of which are described in other chapters, this chapter concentrates on reporting the discussions held on doing these tasks using computers of all sizes. The reporting starts with the use of simple packages, continues with the more sophisticated uses and concludes with a description of international systems which, operating with advanced technology, aim to make bibliographic information available to all.

The Nature of the Data

There are two main kinds of data that can be looked after on the computer; basic data about events, people or objects often called transaction data and master data, and data about data or bibliographic data. Both kinds of data storage and retrieval were discussed in the topic group.

109

Methods of Storing Data on Computers

There are two main ways of acquiring the facility of storing data on a computer; they are to acquire a ready-made piece of software which will do what is required or to build your own package using a programming language. Examples of both were seen.

Prices of packages to store and retrieve data can vary from nothing, they are "bundled" with the computer hardware by the manufacturer, to whatever you can afford to pay. Sometimes professional groups club together to create their own software which is then made available to colleagues at a cost which probably just reflects the cost of the medium storing the software. Sometimes data is included in such packages as well; the example shown by Trisha Strong contained both the manipulating instructions and some data which could be used for teaching.

Building an Information Store Using a Package in Primary Schools

This can be a very simple task on a microcomputer. Trisha Strong described a system that is in use on a BBC microcomputer in the UK, designed for primary school use. This package was developed by The Advisory Unit for Education, in Hertfordshire. The package is called QUEST and is available free to schools. It allows pupils to interrogate some data that has been built into the system and also to build their own datastores.

The example demonstrated how information about horses could be accessed and such questions as "what types of horses exist in India?" answered. The pupil has to learn how to format the question so that the computer can "understand" its task and provide the required answers. Once this technique is mastered the pupil then has the chance to build a datastore more relevant to her or his needs, around any subject or entity that is relevant to the work of the class. This may mean that the computer can act as a tool during a project in the class, enabling the pupil to acquire skills of data capture, data classification and reporting. The use of a computer to do this gives the pupil not only experience of working with computers but the satisfaction of building a useful store of data and information. After some use with a system like this one the student will be ready to move on to a package which allows more data and files to be stored and allows the performance of more complex operations.

Use of Integrated Packages

An example of these was demonstrated by Anne Leeming using a package called Appleworks on an Apple IIe and available from Apple dealers. This package contains three parts, a word processing part, a spreadsheet part and a data storage and retrieval part. It is this latter part that was used at the conference. All three parts of the package can work with data created by the other parts.

She designed a file that enabled participants at the conference to enter their own names and addresses and professional interests. Data could be entered by an individual participant when they were able to do it. The example showed how a screen could be formatted to look like an index card; the names of the data items needing to be entered appeared on the left of the screen leaving the user to enter the values as prompted by the data names. An example of the entries made is shown in Fig. 8.1 below.

```
SURNAME: VEDANAYAGAM
FORENAME: EDITH.G.
PROF. DESIGNATION: PROFESSOR
INSTITUTION NAME: UNIVERSITY OF MADRAS
ADDRESS LINE 1: DEPARTMENT OF EDUCATION
ADDRESS LINE 2: UNIVERSITY OF MADRAS
CITY: MADRAS-600 005
COUNTRY: INDIA
TOPIC GROUP ASSIGNED: LAND WATER MINERALS
TOPIC INTEREST 1: GEOGRAPHIC EDUCATION
TOPIC INTEREST 2: EDUCATIONAL TECHNOLOGY
```

FIG. 8.1. Example of personal data entry.

This package illustrated several useful features of this and similar packages. The initial screen design omitted the line entitled "Prof. Designation" and it was inserted later by request without doing any damage to the data already captured. After the insertion of the new field the package contained spaces for the now missing data item. Later the early records were able to be accessed on the computer and amended to contain the new piece of data.

However, one difficulty appeared with the last two lines of data entitled "TOPIC OF INTEREST". The interpretation of the meaning of these data fields varied from person to person so that some people thought it meant they should enter their hobbies such as cricket or dancing while others entered specific interests within their topic and others indicated different topic groups they would like to keep up with. The point here is that it is sometimes very hard to be unambiguous with the name of a data item especially if the package restricts you to 15 characters for the name of a data item.

This is where cost comes in; a more expensive package will, in general, allow you to be more sophisticated in your data capture by permitting much more flexibility in designing screens. The price paid is not always just in cash; sometimes the capacity of the computer itself is not suitable. There is often a limitation in the size of the microcomputer's memory; moving from an 8 bit computer to a 16 bit computer will allow the use of a package with more extensive facilities.

Data Capture

It is always worth paying a great deal of attention to the design of the data capture process; after all GIGO, or Garbage In, Garbage Out is a well-known maxim. If the data captured for storage within the computer system is not what was intended or is not in the form needed or is not sufficiently accurate then the not inconsiderable labour expended will have been wasted. In other words the quality of the data must be considered when recording it. The design of the data capture process should be rigorously tested during the development stages by getting many different people to try it out. The principles of perception that emerge here are nicely illustrated by the game played in Chapter 2. If data quality is not attended to then the system is likely to have a short life; not being robust enough to withstand many different perceptions of the tasks in the system.

Advantages and Disadvantages of Packages

Using such a simple exercise as creating a name and address file provides a good entry point for people who have not used a computer before. The instructions are easy to follow, the data is known to the user and the task of recording ourselves for posterity is an attractive one!

A further point that was made using this package was how easy it is to manipulate the data once it is stored. The data could be sorted alphabetically by surname, by surname and forename, by country or town, in fact by any of the data items on the screen.

Also queries could be made of the file such as "How many Physics teachers are there at the Conference?" or "How many people in Topic group Food live in Zambia and what are their names and addresses?" Data is easy to add and information is easy to extract from such a file.

However, it must be pointed out that before paying good money for a package care should be taken to see that it does what you want it to; see App 1.5, p.186.

Packages Built by a Programmer

An example of a system built by a programmer was shown by Janis Benkovic. Perhaps more important is that it showed the use of a database of chemical facts by students in the Centre for Chemical Studies in Llubljana. The database contains data about each element in the Periodic Table; a student can access the data on an individual element by attribute, e.g. the oxidation number(s) of the element or its electronic structure. The database could be used to provide precise data given that the user knew the type of data he wanted or to search for elements with given attributes.

The demonstration illustrated the features of a locally built database. One was that it could be tailored exactly to the requirements of the users. Secondly it

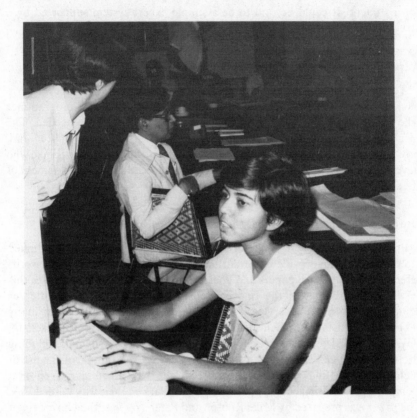

FIG. 8.2. Hands-on experience in the workshop.

would need to have a more developed user interface if it was to be used by a wider user base. The format of the answers to the questions asked needs to be specified and on screen help provided.

A more powerful system has also been developed by Mr Benkovic; this is a system to design a tranquiliser. The system provides basic chemical and physiological data on the constituents needed to formulate a new tranquiliser. The screen shows the two-dimensional structure of the organic molecule and the points where substitution by different radicals can be made. A choice of radicals to substitute is given and when the choice is made by the user the strength of the new tranquiliser is calculated and displayed. Congratulations ensue if a new more powerful drug is found, if not try again! The package combines the use of a structured collection of the properties of the component radicals, some rules for evaluating the new compound and the interactive part for a student. It could be said to be a simple expert system supported by a structured datafile.

Advantages and Disadvantages of Programmer-built Systems

To build systems like these does require more time than to purchase a package directed towards a non-programmer. The services of a programmer who can also become familiar with the chemistry involved are needed. The result is a system whose workings are well known to the originator and, providing the system is well designed and properly documented, the people who have commissioned it will use it. This makes modification and development of the system easier than if it were a package.

Use of the Computer to Provide Data about Information Sources

Professor Bhattacharya described how computers can be linked internationally to provide information on publications; using key words references to published papers and books can be obtained. India accesses this service by mail; though the final search is done on a large computer or computers. Information services such as these have been available for many years, to scientists in the fields of Medicine, Defence, Government research, Physics, Chemistry, Biosciences, Engineering and other related fields including ERIC, Educational Resources Information Centre. An example of the use of such a service is shown in the following extract from a seminar given on a Computer Orientation course for Fulbright scholars on their way to the USA. The paper was prepared by Mike Robson and Peter Towse of the University of Zimbabwe.

DIALOG: a computer search described

During the planning of the ZINFORMATICS Project we wanted to have details of all the articles which had been published in the last 7 or 8 years on computers and information technology in the third world, whether they were originally articles from journals or papers given at conferences and later published.

Faced with such an enormous task where could we start? Well, we could have begun by trying to get hold of as many different journals in the world as we possibly could, at least those likely to contain articles on computers, computing and information technology. We could then have read through, or at least scanned, every issue of these journals for the last 8 years and then built up a suitable list of articles which we had come across.

This is quite a standard procedure but, unless the number of journals is limited, it is a very time-consuming procedure. For example, the attached bibliographies or lists of articles on ecology, the environment and conservation were compiled by one of us in just this way by going through the fifteen journals in the University of Zimbabwe library which publish articles likely to be of use to the secondary school science teacher. The first one, covering the period from 1960 to 1980 inclusive, took well over 200 hours to compile and type. The second, a supplement covering the period 1981 to 1982 inclusive, obviously took rather less.

There are very few journals on computing and information technology in Zimbabwe, so it would have been impossible to follow such an approach in the present case.

However, there are a number of commercial information services overseas, mainly based in the United States, which use computers to carry out a fast search of a much larger number of journals. They charge you a fee which depends on the amount of time the computer is in use. On a visit to Britain in April, Mike Robson used just such a service through the Manchester Polytechnic.

The idea is basically quite simple and depends on the fact that each article listed in the memory bank of the computer is identified by a number of key words or phrases which give some idea of what the article is about.

Let us take a specific example. The article "Small is beautiful (computer development in the third world)" by C. Rose in the September 1980 issue of *Computer* can be summarised as:

Income distribution and employment considerations indicate that, in the foreseeable future, the main users of microcomputers in the third world will be government agencies and the larger corporations. In this article it is suggested that the microcomputer could provide invaluable support to those involved in third world development. However, some fairly formidable problems must be overcome first.

This summary is called an abstract and it is the abstract rather than the whole article which is stored in the computer's memory bank. The reader who has prepared this abstract has considered this article to be concerned with computer development, third world, microcomputers, government agencies and larger corporations. These key phrases are called identifiers and anyone typing into the computer's keyboard one of these identifiers would be able to have access to this article from the computer's memory bank.

All the articles are prepared in this way and so a large file of abstracts and identifiers, called a database, is established in the computer's memory. DIALOG has the largest English language database in the fields of physics, electrotechnology, computers and control. Over 2000 journals are scanned, as are conference proceedings, technical reports, books and university theses.

The enquirer's first task is to decide which identifiers best describe the type of articles he wants. In our case, Mike decided these would be information technology, information revolution, computers, underdeveloped countries, developing countries and third world.

The next step was to telephone DIALOG from the terminal in Manchester Polytechnic's library and connect the telephone to an instrument called a modem, which changes the telephone signals into an electrical form the DIALOG computer could understand.

When contact was signalled, he gained access to the computer by keying into his terminal the appropriate passwords. He then selected the appropriate database and keyed in the identifiers he had decided on. As a result, the computer very quickly found the following numbers of articles:

6	with underdeveloped country or countries in the title
338	with developing country or countries in the title
367	with third world in the title
7	with information revolution in the title
676	with information technology as identifier
33	with underdeveloped country as identifier
790	with developing country as identifier
178816	with information revolution or computer(s) as identifiers

Mike was asked what combination of these identifiers we wanted and responded by calling for those which combined information technology, information revolution or computers with underdeveloped countries, developing countries or third world.

All that remained now was to ask the computer to print out the titles and abstracts of the 92 articles which seemed to meet our needs. That print-out is attached. What comment can you make about the number of articles in this print-out compared with the number of articles which the computer found with the different identifiers?

The search had taken a matter of seconds and the print-out a matter of minutes. Compare that with the time taken to prepare just the listings in the attached bibliographies on ecology, the environment and conservation. Granted the number of references in the bibliographies is very much greater, but you will see immediately the advantages of a computer search such as this.

We still needed to obtain the original articles, of course, for the computer stored in its memory only the abstracts. Very few of the 92 on the print-out were in the University of Zimbabwe library. However, we were able to order copies of the others through the Inter-Library Loan System, which allows any library to ask other libraries for the loan of books or journals which they have in their stocks, or for photocopies of appropriate articles. This system operates throughout the world. The first article to be received came from a library in South Africa and is attached. Do you think the abstract provided by DIALOG is a fair summary of Nwachukwu's article?

Since one cannot operate this search system directly from within Zimbabwe, it is impossible for you to perform a similar one on this course. However, such searches are commonplace in the United States and you will encounter this routine at the university to which you will be going.

9

Packages—Learning at a Distance

JENNY PREECE AND ANN JONES

Introduction

Distance Learning is a form of learning which has the following characteristics:

1. Unlike the traditional classroom teaching situation, the learner and the teacher are not usually located in the same place.
2. Most learning is through correspondence texts, but radio, television, video, audio-cassette, computers and home experiment kits may also be used.
3. The responsibility for learning rests with the learner, who is usually an adult living at home with a full-time job.
4. The teaching materials are usually produced in a large scale systematic way, using methods which are similar to those of a publishing house.

The following description of the British Open University illustrates some of the features of distance learning.

A Case Study of Distance Learning—the Open University
(G. Hollister and Jenny Preece)

Background

The Open University like all British Universities, is an autonomous institution with the freedom to design its own curricula. It began in 1971, teaching adult students (minimum age 21) to first degree level in six faculties: arts, educational studies, mathematics, science, social science and technology. In addition, the University has an Institute of Educational Technology, whose members are concerned with the design, development, monitoring and assessment of all the courses produced.

Students attain an ordinary degree by accumulating six course credits, or eight credits for an honours degree. A full credit consists of 32 or 34 weekly

117

"units" of work. Courses are multi media: each unit of work usually contains a correspondence text, together with assessment material, an associated radio and television programme and, for science and technology courses, some of the units are accompanied by a home experiment kit. In addition, there may also be cassette or disc recordings, models, film strips, computer assisted learning, etc. Each unit requires approximately 12 to 15 hours' study time, and the diagram below shows how this time is allocated. To obtain a credit a student has to perform satisfactorily both on continuous assessment (regular tutor and computer marked tests) and a final examination. For several courses the student is expected to attend a one-week "summer school" on the campus of a host university where, for science and technology courses, the student is subjected to fairly intensive laboratory work.

Perhaps the most salient feature of the Open University is that most of its students are combining their study with their everyday working lives and family commitments by studying at home. Students come from most occupational categories and there are equal numbers of men and women.

At present (1985) there are about 100,000 students registered on OU courses. Two types of courses are available; undergraduate courses which form the first degree programme of study and Continuing Education courses. Continuing Education courses cover a wide variety of areas from management and professional and scientific updating to short general interest courses on subjects such as child rearing or saving fuel costs. Examples of two such courses are given in the next section.

One in twelve British graduates is now emerging from the Open University, and so far over 40,000 adults have graduated through part-time study. Each year there are 60,000 applications of which 30–40,000 are admitted. Most encouraging of all, the crude cost of each student per annum is modest compared with at least five times that amount at a traditional British university.

The Open University is characterised by the following unique combination of features:

Students study part-time from home.
An open admissions policy.
Courses are multi-media.
Specially designed "Foundation Courses".
The use of teaching aims and learning objectives as a means of course specification, communication and assessment.
"Course teams" create the courses.

The implications of some of these features are discussed below.

Foundation Courses for an Open Admissions Policy

The Open University has been committed to a policy of open admissions from its inception. What are the implications of an open admissions policy? Available evidence confirms that to open admission to less qualified students without making any effort to change teaching patterns accordingly is both unwise and, ultimately, very unkind. At the Open University the solution to this problem was the creation of foundation courses specifically designed for students about whose educational preparedness we could hold no preconceived notions. For example, the technology faculty's first foundation course, "The Man-Made World" (T100), as well as its replacement, "Living with Technology" (T101), assumes nothing about the student's capabilities except that the student can read, write, add, subtract, multiply and divide. To cater for varying degrees of skill and knowledge, there are alternative routes through the course for students with different skills, and special "tributaries" of additional material (e.g. in Maths) for students who may need it.

One of the reasons for the relatively high success rate of OU students is the great attention that university staff have given to creating courses specifically designed to ease adult, often unqualified students into higher education.

Teaching Aims and Learning Objectives

When teaching at a distance, via correspondence, radio and television, it is important to be clear about the precise nature of what is to be taught, the depth to which it is to be taught, and how it is to be assessed. Most courses and units therefore contain teaching aims—a broad statement of what the teach intends to achieve; and the learning objectives—a precise statement of what the student should be able to demonstrate about what he has absorbed.

In order to help students assess for themselves whether they have properly achieved a given learning objective, Open University course units are also liberally interspersed with self-assessment questions. Thus, OU course units are written in a manner which conforms closely to the age-old teaching principle:

1. Tell them what you are going to tell them (state the learning objectives).

2. Tell them.

3. Tell them what you have told them (or, in the case of OU material, ask them to check for themselves what they have learned).

Aims and objectives fulfil two other vital functions: they help the course team to decide which objectives are best achieved by which means (correspondence, radio, TV, summer school activity, home experiments, etc); and they are invaluable in writing examinations and assessment material.

Correspondence Texts and the Course Team

Open University courses are produced by a "course team" which consists of a number of different specialists: academics, a BBC radio and television producer, an educational technologist, a course manager, and (for foundation courses) the faculty editor.

Typically the materials go through three drafts, each of which is circulated for comment to members of the course team and experts in other institutions. In addition, many courses are developmentally tested with students so that any logistical or conceptual problems can be identified and rectified before the materials go to print.

As all the above tasks have to be undertaken to rigid deadlines, they can be onerous and exhausting. Furthermore, academics are not used to having their material criticised by colleagues. Nevertheless, the system works well and the quality of the resulting product is invariably high, as indicated by the growing use of OU material in teaching establishments throughout the world.

The Role of Radio, Television and Video

Broadcasting accounts for about 22% of total OU costs, although transmission costs account for only 5–6% of total broadcasting costs. Does the university get good value for such a huge expenditure? The strongest argument is that broadcasting provides experiences which are, at the moment, impossible for our students to achieve in any other way, and that sometimes without these experiences, the student would not be recognised by other institutions or other universities as having reached a "degree" standard. In science and technology, therefore, television is used for bringing to students experience of experimental situations, which they could not otherwise have. Television and video are used in the following main roles:

(a) To illustrate dynamic phenomena which cannot be put across by other means.

(b) To demonstrate particular experiments.

(c) To show research, development and industrial work and facilities.

(d) To interview eminent experts and specialists in connection with subjects relevant to the course.

(e) To provide "primary source material" e.g. to show sequences taken in distant parts of the world.

Radio is also frequently used, for instance to interview experts, as with TV, as a back-up role to the TV programme, to "talk students through" particular points, or to broadcast tutor-student discussions.

Teaching Science and Technology

Most science and engineering courses require a degree of laboratory work. In deciding which experiments students should undertake, OU course teams must state clearly, in terms of behavioural objectives, what the student is expected to be able to do or understand as a result of conducting the experiment that he could not do or understand before, and explain why these objectives cannot be met by other means. Having decided that an experiment is essential, the following possibilities are open:

(a) An experiment conducted on television with the students participating by observation, data collection and subsequent analysis (data sheets would be provided with the course material, and the students' analysis assessed by a tutor).

(b) The development by the course team of a low cost home experiment kit so that the student can conduct an experiment at home. Many such kits have been developed, including microscopes, fluid dynamics experiments for use in the kitchen sink, plastics for polymer engineering, materials testing kits with photoelastic polariscope, management system "games", miniature CROs for use in circuit design, microcomputers, purpose-built microchips, and many other devices. Many of these kits are in great demand with other teaching institutions.

(c) Traditional laboratory work carried out during the summer school in the laboratories of a host university.

By the above means, the needs of OU science and technology students for effective laboratory experience are, we believe, adequately met.

Distribution and Regional Support

Students receive their course materials through the post—all in one package for some short courses and at intervals throughout the year for longer courses. Clearly for such a system to operate there must be a reliable postal system—and as we shall see later in the chapter for many developing countries the lack of such a reliable system means that this is not a viable option for them.

Students also use the post to submit regular assignments to their tutors, which are one means of assessment: others include computer marked assignments marked centrally, and exams. The majority of assignments are sent to and marked by local tutors.

The university employs some 5,000 part-time staff as local tutors and tutor counsellors; many of whom are lecturers with other institutions or colleges. They mark and grade the students' assignments—developing a dialogue by writing comments on the scripts, with their grades, before returning them to the student. Tutors also organise and run tutorials and discussion groups at study centres, of which there are 250 scattered throughout the United Kingdom. All these activities are co-ordinated by the full-time staff of 13 regional offices, who employ, train and supervise 5,000 part-time staff and organise study centre activities.

Two Examples of OU Materials (contributed by Ann Jones)

As has been explained earlier, the Open University has two main sectors, the Undergraduate sector and Continuing Education. Both adopt a multi media approach but the major differences are as follows. Naturally for the undergraduate programme each course is part of a degree profile, it is always assessed, has tutorial support and is studied within a set time period running from February to October: a full credit is expected to take an average student 12–14 hours study each week throughout that period. Continuing Education courses on the other hand may also be assessed and studied over a fixed period—and may have some tuition or they may be studied at any time as totally stand alone packs. The study time may be anything from a few hours to a year, and the level anything up to postgraduate. In the Workshop we used two packs as examples.

Example 1.   Health and Productivity in Dairy Cattle

The "Health and Productivity in Dairy Cattle" pack was chosen because it is a practical pack, produced in collaboration with the farming and dairying industry, and because its contents are accessible to a wide audience including developing countries.

It is designed to assist dairy farmers and stockmen increase their efficiency and profitability. Like many other packs it is multi media, consisting of:

4 TV programmes
2 Workbooks (the main study text)
Audio Cassette
Case studies
Information sheets
3 computer marked assignments.

The course runs for 12 weeks taking 6–7 hours a week and local tutoring is provided—often the local veterinary surgeon.

The second example is the micros in schools project.

Example 2.   Micros in Schools Packs

Some of the Continuing Education packs provide inservice training materials for schoolteachers. The recent and extremely rapid development in microcomputers and microelectronic devices and their introduction into schools has led to a great need for inservice teacher training. The Open University has developed five packs covering the following aspects of inservice teacher training:

1.   Awareness training in which the aim is to develop teachers' confidence and competence in using microcomputers. (1 pack).

2.   Second level training packs to help teachers make the best use of educational software. (2 packs).

3.   Training in basic microelectronics and the fundamentals of computer hardware. (2 packs).

We believe that it is important to design materials which are relevant to teachers' needs and which build upon their existing experiences and consequently some of our packs may not be suitable for use in the Third World as they stand. However, there are many reasons why Third World countries may wish to consider training through distance teaching as opposed to conventional training—for example, the problem of reaching scattered and isolated students and the lack of qualified teachers. Within this context, the fundamental design philosophies developed by the OU and exemplified in these packs are relevant.

The History of the Micros in Schools Project

The Open University Micros in Schools project started in 1981 and was funded by the Microelectronics Education Programme (MEP); itself funded by the Government.

The project initially produced an awareness pack which was successful and following a further grant of £486,000 in 1982 produced four more packs on the topics of "educational software", "microcomputers in action in the classroom", "microelectronics", and "inside microcomputers". The first two packs named above were to be produced in editions for the Apple II, Sinclair Spectrum, BBC model B, Research Machines Ltd., 3807 and LINK 480Z. The microelectronics and microcomputers packs were to be supported by a specially designed, battery powered, microcomputer named by the acronym DESMOND (digital electronic system made of nifty devices.) The complete range of packs was available in early 1985 and will remain on sale until at least 1987. All the packs are designed so that they are suitable for teachers of all subjects. This is done by including a wide range of case studies—"something for everybody"—and by encouraging teachers to relate principles to their own teaching situations. In the workshop three of the packs were shown, "awareness", "educational software" and "microelectronics".

Pack Production

All the packs provide the support necessary for teachers learning on their own; for example, key presses are identified, and likely error messages are explained and corrected. An important design feature is the three column format used to describe the computer activities—this is what you do; this is what you will see; and this is why it happened. Every effort is made to ensure that the materials are relevant, easy to use and that the software is clearly explained and bug-free. In order to achieve this quality, each draft pack is developmentally tested in draft form with real teachers in real conditions. The test is run by an educational technologist who is independent of the team producing the pack, and who finds out whether testers get stuck with the material and what intervention, if necessary, is appropriate and to report back to the project team. Novices encounter a whole variety of conceptual and practical problems with computing work and there is no way of determining a priori what these problems will be, or the best way of, say, explaining new ideas like assignment. The strategy adopted by our project has been to continue testing material until it works. Apart from the usual course team the staff include programmers, an information officer, a market researcher and sales staff.

Each pack took between 18 months–2 years to produce. The main components of the packs, like most other OU courses are:

Study guide/text
practical work   — giving hands-on experience
case studies   — documenting existing practice
video   — to illustrate and provide shared experience
audio   — to provide shared experience
software   — provide programs for teachers to use
Reader   (in Educational software)

At the final stage, just prior to printing the page proofs are checked against the computing practical work; a single wrong key press could cause problems for students who are learning on their own.

The Awareness Pack

The Awareness pack is designed for teachers who have no knowledge or expertise in using microcomputers in the classroom. The aims, as stated in the pack are to train teachers to:

connect up, switch on and run a microcomputer;
use educational material on the microcomputer;
evaluate the educational potential of the microcomputer;
understand enough computer jargon to be able to communicate with a local computer expert.

We believed that the best approach for achieving these aims was through the knowledge and skills that the teachers had acquired in their professional training and experience. In order to capitalise upon the teachers' expertise, we realised, at a very early stage, that the teachers needed to be helped to acquire confidence and competence to use the new technology effectively, and that the only way to do this was through extensive "hands-on" experience. The three column format, boxed key presses, and extensive trouble-shooting guide were, therefore, developed to provide a high degree of "hand-holding" in order to facilitate the teachers' "development" of the basic skills, knowledge and confidence needed to use the equipment. Like all micros in schools packs the Awareness pack has a study book which directs the teacher through the pack and introduces basic concepts and also introduces the first hands-on activities including two long projects based on a queuing simulation and an information retrieval program. Software disks are also provided and short case studies written by practising teachers.

In essence therefore, the role of the Awareness pack is to take teachers from a basis of "no knowledge" and to provide a foundation upon which to develop the skills and knowledge necessary to select software and to plan and evaluate its use in the classroom.

Educational Software Pack

"Educational Software" is designed to give some insights into the nature of educational software and how it is produced.

Its aims include:

providing an appreciation of fundamental programming concepts,

introducing some principles underlying educational software design,

helping to develop an appreciation of which kinds of software are suited to which learning tasks.

The only prerequisite for this pack is to have studied the Awareness pack or have equivalent experience.

Briefly, the route through the Educational Software pack is as follows:

1. Introduction.
2. Case study—looking in detail at one educational program.
3. Programming—how educational software is designed and written introducing programming concepts—(mainly using Logo).
4. Styles of educational software—introducing different kinds of educational programs, how they're written, and what they're used for.
5. Selecting Educational Software—developing criteria for selecting educational software, by looking at commercial software.
6. Developing educational software—by examining how educational software is produced in Britain.

The "Learning about Microelectronics" and "Inside Microcomputers" Packs

Both these packs are designed to give the teacher a greater understanding of the machines and devices which constitute the "new" technology. "Inside Microcomputers" teaches about the architecture and components of computer systems—the nature of computer processors, memory, interfacing and some programming at the machine level. The aim of the Microelectronics pack, which was shown at the workshop, is to develop some of the basic ideas about digital microelectronics, such as the nature of digital signals, the types of devices which produce them and respond to them and the types of devices, or combinations of devices, which can be used to manipulate them.

The common feature of the two packs is DESMOND, the special computer mentioned earlier in this chapter. DESMOND is programmed using a simple calculator style keyboard with a liquid crystal display. It also includes a variety of sensors for light, heat, touch, magnetism and tilt, some of which are simple ON/OFF switches and others of which produce an analogue signal; there is also a similar range of output devices—a motor, buzzer and lights (both digital and analogue). A simple switch converts it from a specialised board for use with the microelectronics pack to a simple microcomputer to accompany the microcomputer pack. DESMOND is used in different ways in the two packs but allows both packs to have a substantial amount of practical activity including using input and output devices, other than the keyboard and screen, to extend the student's experience.

In the Microelectronics pack DESMOND acts as a way of connecting the input and output devices together. All the connections are made by a few keystrokes and there is no soldering! The student can also incorporate combinational elements, such as logic gates, in "circuits" to produce, for example, an electronic spirit level, burglar alarms, times and so on. The teachers' circuits are emulations within the DESMOND system so they are easy to create, amend and correct without any of the practical problems inherent in constructing them from real components; the teacher is, therefore, free to concentrate on the concepts involved.

Despite the simplicity of the system, the practical work leads the teacher to quite sophisticated problems, using the input and output devices. For example, during testing in a school DESMOND was made to control a model automatic washing machine. Most of the fundamentals of larger computer systems and processor designs are accurately represented within DESMOND.

The components of the two packs are similar:

The Practical Book of DESMOND activities,
An Illustrated Book which gives a theoretical background,
Study Book to suggest routes through the course,
Case Studies to show the way teachers have tackled the problems of teaching these subjects,
Video and Notes which examine in detail the effect of DESMOND in two schools.

Conclusions and Possible Implications for the Third World

We have briefly discussed some inservice training packs provided by the Open University Micros in Schools Project and have also described our pack production philosophy. The main thrust of awareness training is towards competence and confidence, whereas the main functions of the second level courses has been to help teachers to develop sound frameworks from which to make informed judgements. We, as trainers, have adopted the role of "question raisers" rather than attempting to provide answers. In order to do this, we have had to search for the right balance between introducing and discussing educational perceptives (theory) and building on teachers' own experiences (practice). We have also sought to provide thoroughly tested materials with "bug-free" software and the "hand-holding" support that teachers need in order to develop their own frameworks.

In our discussion of the packs we stressed the importance of introducing new concepts by building on teachers' already existing experience. One way that we have achieved this is through case studies of classroom practice and consequently some packs are strongly geared to British schoolteachers. Of course, any Third World country considering embarking on this route would want to gather case study material relevant to its own situation. Another solution to this problem would be to design materials specifically for the Third World. However, the cost of producing these kinds of packs is very high. Costs could be cut by reducing the amount of testing of drafts or by reducing the amount of support provided by specifying each "key-stroke". Such economics would, however, result in much poorer products which would undoubtedly not fulfil the main aims of helping teachers to develop competence to use microcomputers and microelectronic devices. Another problem that has drained our project's resources is maintenance. New microcomputers come on to the market and already existing models are sold with new operating systems or graphics capabilities which require us to modify our software.

One of the conclusions that our project has reached is that producing high quality relevant and well supported training packs is expensive and cutting corners to save money is not possible without a considerable reduction in quality. Most of the packs described in this paper would be suitable for use in Third World countries, if relevant case study material were used. The exception is the Microcomputers in Action in the Classroom because it focuses on British classroom practice.

The Relevance of Distance Learning to the Developing World

The following data contributed by Jenny Preece shows the presence of distance learning institutions in different regions of the world. The figures are taken from a survey which was carried out in 1983 and reported in a publication by Lord Perry (1984), the Honorary Director of the "International Centre for Distance Learning of the United Nations University", and ex Vice Chancellor of the British OU.

Region	No. of countries approached	No. of instituions engaged in DL	No. institutions founded for DL
Africa	22	16	10
Asia	13	25	6
Australasia	4	40	15
E. Europe	6	0	0
W. Europe	17	101	57
Middle East	3	1	1
N. America	2	102	30
S. and C. America	21	19	7

These figures show that although the greatest number of institutions engaged in distance learning and particularly those that were funded especially for distance learning, are in Western Europe and North America. There are also significant numbers in Africa, Asia and Australasia.

It was pointed out that the Asian institutions are particularly impressive in terms of the number of students studying with them. The average number being 55,000 students per institution. Thailand and Korea were mentioned as examples of two very large institutions which at that time (1983) each had over 120,000 students. Today (1985) there are about 200,000 students taking courses with Sukhothai Thammathirat Open University (STOU) in Thailand. Many more interesting statistics could have been discussed. The purpose of mentioning these facts was to demonstrate to the delegates that distance learning in the developing world was not just a speculative suggestion made by delegates from the UK Open University, it is a practical solution which is already being successfully operated in several countries in the developing world.

Two questions need to be considered:

Why set up a distance learning system?

What kind of support is needed to start a distance learning institution?

The first question can in part be answered with the following quote from Professor Wichit Srisa-an (1981), the Doctor of Sukhothai Thammathirat Open University in Thailand:

> In developing countries opportunities for education in the traditional system are somewhat limited. Since the level of economic and social development of a society is closely related to its stock of values, attitudes, knowledge and skills, both productive and social, it is essential to have a teaching mode that will enable a more extensive and egalitarian basis without having to stay away from their jobs to attend classes. The distance-learning system can thus be seen as an effective and economic means of extending educational opportunities.

The possible advantages raised in Dr Wichit Srisa-an's quote can be summarised as:

1. It is cheap compared to conventional university education.
2. Distance Learning reaches a large number of people over a wide geographical area.
3. Distance Learning provides adults with education whilst they continue with their ordinary everyday jobs.

The second question is difficult to answer directly but the following points need to be taken in account:

1. Political acceptance and support;
2. the necessary financial support, and
3. an adequate communications system.

In the UK and most of the Western World the main form of communications is usually the postal service, but in other parts of the world other systems may be used instead of the postal system which is often unreliable. In the West Indies, for example, satellite is extensively used to communicate between the islands. Radio is also an important alternative in many developing countries if a well-developed network has been established. Television is another alternative but is often more expensive. In the UK, for example, television costs about seven times as much as radio.

Expertise and technical assistance are also essential to:

(a) develop and run the programme;

(b) write, develop, test, disseminate, update and maintain materials.

The important thing to consider when thinking about these questions is that every country is different and has its own strengths and weaknesses which will influence the answers.

The University of West Indies Distance Teaching Experiment (UWIDITE) (contributed by Vilma McClenan).

The University of West Indies Distance Teaching Experiment (UWIDITE) links the three University Campuses in Barbados, Jamaica and Trinidad and the Extra Mural Centres in Antigua, Dominica and St Vincent via a leased telecommunications network. Jamaica and Trinidad are linked via a leased telecommunications network. The way that UWIDITE operates is quite different from that of the Open University in the UK.

Communications

The communications systems are excellent when they work but there are many problems, but before discussing these problems, Vilma outlined how the system worked.

Communications are carried out in various ways:

Jamaica and Trinidad are linked by INTELSAT satellite, while the Eastern Caribbean countries—Antigua, Barbados, Dominica, St Lucia and Trinidad—are linked by microwave, UHF and tropospheric scatter. The network is bridged in St Lucia.

Each UWIDITE centre is linked to the international gateway (JAMINTEL, TEXTEL, BTEL, and Cable and Wireless) by leased four-wire telephone lines.

Each centre is equipped with microphones, speakers, convenor, telephone handset, switching unit, slow-scan television system, and telewriter for on-line use. Communicating computers are also available at each Campus site. Audio-tapes and Video Cassette Recorders and Monitors are also available for off-line use at all sites. Teleconferencing is also used to provide two-way communication between one or more people at various sites remote from one another, using electronic means.

Distance Teaching

Teaching originates mainly from the campuses with students located at some or all sites. The courses offered include:

Principles of Training for Day Care Personnel
Certificate of Education for Teachers of the Hearing Impaired
Certificate of Education for Teachers of Reading
Certificate of Education for Teachers of Mathematics
Secondary School Laboratory Technicians Upgrading
BSc Part I Social Science Courses
Continuing Education for Doctors and Senior Nurses in Reproductive Health
Law Tutorials
Nutrition for Community Health Workers

"One-shot" programmes, e.g. Discussion series on Caribbean Art Forms, Creative Writers Workshop, Science Quiz are offered and the Departments of Obstetrics and Gynaecology in all five countries meet monthly to discuss case histories and patient management.

Programmes planned for the future are:

Agriculture
Energy Management
Disaster Preparedness
Computing

UWIDITE Objectives

To demonstrate that a sufficient level of actual demand exists to support an operational system.
To create within the University of the West Indies a desire to meet identified demands by teleconferencing techniques, so that the institution will incorporate these as part of its armoury.
To establish a core of experience and experienced workers to allow the efficient design, staffing and implementation of an operational system for distance teaching and outreach.
To help develop mechanisms and expertise for the production of educational material, for example, print, audio and audio-visual material.
To develop a proposal for a permanent service.

Funding

USAID grant of US$600,000 over 3 years (May 1982–May 1985) to cover salaries, travel and transportation, preparation of educational materials and other direct costs.
USAID also funds the rental of the telephone land lines and international

circuits, purchase of audio equipment and technical assistance contracts with the Academy for Educational Development, Teleconsult and Abt Associates Inc., all based in the USA and assisting with training, technical matters and evaluation, respectively.

Commonwealth Association of Science, Technology and Mathematics Educators (CASTME) provided $99,000.00 for the teaching of Science Technicians in the Caribbean over a 3-year period (Jan. 1983–Dec. 1985).

Additional funding for visual equipment has been provided by European Development Fund (EDF) and Johns Hopkins' Programme for International Education in Gynaecology and Obstetrics (JHPIEGO).

Various other organisations support programmes which are of their own particular interest.

Some Problems

As has already been said the system of communication is thus quite different to that of the British Open University, and when the telephone system fails the centres may be cut off for quite long periods of time. Another problem is that the cost of renting the INTELSAT for 5–6 hours per day is high. There is a mail system which is used to distribute texts but this is often unreliable and is made worse by the authors of the texts not submitting them on time for printing and distribution. Power cuts and problems with telephone lines cause frequent interference and breaks in broadcast transmissions.

Distance Education in Nigeria (contributed by Dr E. I. Alonge)

Distance education started in Nigeria in 1974. The main features of the system are that:

1. It is based in a conventional university.

2. It runs undergraduate courses.

3. It utilises full-time faculty staff to write the course texts, teach at the centres and broadcast programmes.

4. Students train/learn whilst doing their ordinary jobs.

5. The cost of studying this way is lower than in conventional universities.

6. The degree programme lasts for 5 years.

The students are helped by the provision of some tutorial support. Other universities act as study centres and the students meet fortnightly and for longer periods during the long summer vacation.

The system works quite well, but not surprisingly there are some problems. Full-time academic staff are over-stretched: the foundation courses were very difficult to prepare; there are some organisational problems; and the regular university structure sometimes resists the demands which are peculiar to open learning.

Computing in Developing Countries (contributed by Gordon Davies)

In this final contribution the importance of computing to developing countries is considered, with particular attention to their particular training needs and how these might be met through distance learning.

Importance of Computing to Developing Countries

It is no exaggeration to indicate that organisation of computing technology transfer is one of the most critical variables facing developing countries.

A. El Sayed Noor, (1984)

Bearing in mind the present trends in computing technology and applications, computers have a significant role to play in social, economic and political development.

A. El Sayed Noor, (1984)

Computer education in a developing country can be misguided. A great deal of effort may be spent on teaching a high level language, as though the sole objective of computer education was to obtain some proficiency in programming. Development countries often regard the appearance of computers as a sign of progress, they require the most advanced machine without regard to the usefulness of such a purchase. This poses problems. For example, a government department that is discovering how to use a photocopier may suddenly be required to make use of a large prestigious computer. The use of computers in centralising power within any form of government is another major issue, although this problem is not dealt with here.

Such an important subject as computing requires adequate education and training. Although the shortage of qualified staff is not unique to computing, the profound economic and social impacts of computer technology warrant greater concern. This is also being realised by the developing countries such as USA, France, Japan and UK, who are investing more in computing. In the UK the investment is not only in research but in providing training in information technology. Rather belatedly it has been realised that there is a shortage of qualified people in this area. This was compounded by the shortage of suitable training courses.

What should be taught?

Unesco list 4 levels of computer activity in their document "The application of computer technology for development" Pub No E71.11.A1 United Nations, New York.

Initial — no operational computers
contact via salesmen.

Basic — some understanding of computers in government and private decision centres.
A few computer installations. Some nationals are involved. Some government use.

Operational — extensive understanding in government and business. Centres of education offering degree programs in Computer Science. Design and production of software. Some manufacturing

Advanced — most government work is carried out by computers. Well established activities. Complete range of technology education.

The Basic and Initial groups provide most of the foreign students entering US universities. There are few trained nationals (Horowitz, 1977).

Many of these overseas students have no computer science training in their undergraduate programme and take pseudo postgraduate courses (conversion courses) in UK or amalgamations of UG courses in USA.

It is believed that developing countries do not need technicians only, nor do they need one-sided specialists. Graduate Computer Scientists returning to their own country from overseas training have to deal with many aspects of computing. Their training should prepare them for this.

One can summarise three aspects of computer education.

(a) Secondary—to enable efficient use of computers in someone's own speciality.
E.g. teachers,
secretaries using word-processors efficiently,
research scientists.

(b) Vocational training and computer systems education—to provide programmers/analysts for industry.

(c) Computer science education—to advance the state of the art in theories and applications. Universities have responsibilities not only to train but to educate and present state-of-the-art and future developments. Tomorrow's developments depend on this part of university education.

As already stated a graduate returning to a developing country may have to be skilled in all of these activities.

What then should we be trying to give to the developing country student? After a computing education he should be better equipped to tackle the following:

(a) determining when automation offers a feasible solution or an improvement;

(b) convincing policymakers that computerisation is appropriate;

(c) formulating systems specifications and requirements;

(d) writing or contracting software;

(e) training personnel to use a system.

(Horowitz, 1977)

What then should go into a course for developing countries? To provide training in the above five areas a special course should be provided. It should be an integrated course from Computer Science (Software), Electronic Engineering (Hardware), and the Business School (Business Context Economics).

This is more appropriate than a single department based course. There are four components to such a course.

1. Hardware — technology as it is, the range of computing equipment, networks.

2. Software — programming and then software design and project management. Communication with systems.

3. Economic — costs of systems.

4. Social — applications in various areas of society—banking, health care, industrial processes.
   — effect on people.
   — retraining.

How can distance learning help?

Education and training is a major source of a long-term supply of skilled human resources and investing sufficiently in the education and training of computing personnel is important.

The currently available computing courses are a mixture of postgraduate and undergraduate courses. There is little computer science in developing countries although the situation has improved in recent years. A criticism of current courses in the West is that they prepare students for a Western society and that there is the inevitable loss of developing country students to the developed countries. The student's expectations are higher after being educated in the West and he often prefers to stay in the West to improve his knowledge as well as his financial situation and experience.

The evolutionary path followed by universities and academic institutions in developing countries has been criticised in these countries. Following their example might not be the best way to achieve the national computing needs of developing countries.

Students from developing countries attend courses geared to the West, whereas they would be better served by attending courses with a curriculum particular to their needs. An alternative to sending students abroad or providing high cost education in their own country is to educate them in their own country by Distance Learning at much less cost. The student need not leave his country or his job and one hopes that the problems of developing country students staying on in the West will be diminished.

Courses can be adapted to the individual country's needs and are not developed from scratch. As with all distance learning material, the costs are less because of this and their re-usability.

The OU has already become involved in providing computing courses at both undergraduate and postgraduate level and the latter courses are aimed at retraining graduates, or their near equivalent, while the student is working. They are also designed to update a student's knowledge in computing, assuming very little basic knowledge.

Such courses could be useful in a developing country where computing staff have little formal training and are expected to acquire skills while working. A common situation is that of a maths graduate who is employed as a programmer. Typically such people are now trained in the West. Our experience with retraining courses suggests that such technical staff can be trained in their home country while remaining in post, at considerably less cost to the student, the company and the country.

Additionally there is the situation of a graduate or equivalent working in an engineering or business concern who finds that computers are becoming more important in his job. That is, his job has become increasingly dependent on the use of computers. If his knowledge of computers is increased he will be a more able and productive employee. Thus at the higher levels of technical staff, and management too, retraining can be done at home while staying with the company, with the advantages previously mentioned.

Comment

Dr Ploman, a vice Rector of the United Nations University, has commented that a multi-disciplinary approach is essential and that new technology should be used when relevant. It is also important for countries to share materials and expertise, but the specific needs of developing countries should be taken into account and, where possible, developing and developed countries should work together. Some rethinking will be necessary in the light of the present information technology explosion. It is also essential that open universities and traditional universities should work together in a complementary way and that they should not compete with each other. Distance education will be important for reaching a large number of people over a wide geographical area and in countries such as India where the population is very large. In the case of India the distance learning system now caters for 3 million students, and in China a system is being developed using television and other communication techniques, which will provide education for very large numbers of students.

References

Hollister, G. S. (1985) Distance Learning Systems. Paper No. 14 for the Information Transfer and Technology Topic. Science and Technology Education and Future Human Needs. Bangalore Conference: ICSU/CTS.

Horrowitz, E. (1977) Training Computer Scientists for Developing Nations. *Communications of the ACM*, **20** (12), 968–970.

Noor, A. F. (1984) A Framework for the Creation and Management of National Computing Strategies in Developing Countries. *The Computer Journal* **27** (3), 193–200.

Perry, W. (1984) The State of Distance Learning Worldwide: The first report on the index of institutions involved in distance-learning. International Centre for Distance Learning of the United Nations University, c/o The Open University, Walton Hall, Milton Keynes, UK.

Srisa-an, W. (1981) The education of adults at a distance: an Asian perspective in Neil, M. W. (ed) *Education of Adults at a Distance*. Kogan Page (London), pp 23–27.

10

Teaching about Information Technology in Schools

MIKE ROBSON

TEACHING about information technology as a subject in its own right, as opposed to using it as a medium of instruction, is a very recent development. But it is clearly going to play an increasing part in science and technology education as the world comes to terms with the explosive increase in the use of the various new technologies. Examples of teaching programmes from four countries with very different backgrounds and needs will show how far the subject has progressed.

We will begin with part of a paper from Israel which not only describes a particular course, but also sets the scene. It is entitled "Man coping within an Information Society" and is by D. Mioduser, D. Chen and R. Nachmias.

MAN COPING WITHIN AN INFORMATION SOCIETY
D. MIODUSER, D. CHEN and R. NACHMIAS
Tel Aviv University, Israel

The contemporary period clearly exemplifies the impact of technological innovation on both individual and society (McLuhan, 1964; Bell, 1979; Forester, 1980). It has been termed Information Society (Bell, 1979); The Electronic Age (McLuhan, 1967); The Third Information Revolution (Simon, 1977), and The Computer Age (Dertouzos, 1979). The common denominator of all these terms is Information Technology.

Unlike technological innovations in the past, current technology is changing very fast (Abelson and Hammond, 1977). It has only been 40 years since the introduction of the first electronic computer. Today information technology is everywhere—industry, services, defence, entertainment and home. The enormous changes in both individual and social institutions impose heavy demands on the new types of knowledge and skills, which must become an integrated part of the individual's basic education.

Being able to cope with information in this environment requires the ability to select, store and retrieve required information. The new jobs created by the Information Industries require vocational education which did not exist in the past. Adaptation to changes in the mode of operation of traditional social institutions requires understanding and relating to these changes which depend on developing pertinent attitudes and skills. The availability of the technology to the individual, in his immediate environment, requires careful handling and mastery, in order to prevent the abuse and misuse of these powerful tools.

137

The need for the traditional curriculum to change and adapt to these new needs is widely recognised, and its importance is emphasised by a series of reports and papers (Licklider, 1979, Klassen, 1981, Chen and Nachmias, 1983). Nevertheless, educational institutions move very slowly, compared to the speed with which information technology is changing and moulding our lives.

In the late seventies, the feasibility of using information technology for educational purposes has created, for the first time, an awareness of the need for introducing new elements into the formal curriculum. Such elements should enable the wide-scale preparation of future citizens to life in the "information age".

Two major levels of teaching were usually adopted: the first involved introducing computer science as a subject, mainly in high schools and vocational education. The other involved introducing it as "computer literacy", which would enable the entire population of students to interact with computers adequately. It is this second approach which interests us in the present paper. There are different versions with regard to the scope and sequence of computer literacy courses (Chen and Nachmias, 1983). According to Luehrman's definition (1980), computer literacy is intended to give the student "the control over the computer", namely, the skills and knowledge necessary to use it freely. While we feel that this is probably an indispensable part of modern education, it is most definitely not a sufficient measure to enable the graduate of schools in the 1990s, to meet the demands of the information age.

The present paper describes the concept of "information literacy", as the main framework within which the concept of computer literacy may be defined. We will describe in detail the design of a complete curriculum in information literacy, intended for all students in elementary and junior high schools, and we shall illustrate this design with a specific learning unit, titled "Information in-Formation".

This curriculum, which was developed by the Tel Aviv University Science Teaching Center, is a part of a larger program, titled "MABAT" (the Hebrew acronym for Science in a Technological Society), which will be integrated as a part of the national science curriculum in 1985.

I. The Design of the Course: Man Coping Within an Information Society

Central to the project "Science in a Technological Society" is the relationship between man and environment, emphasising the extension of man's natural capabilities via technology (Chen and Novik, 1984). "Man coping with Information Society" is a major segment of the above curriculum. It is intended for all students in the educational system, which enables its inclusion in the curriculum of compulsory education (grades 1 to 9).

The objectives of the program are:
1. Understanding the very nature of information as a human and technological phenomenon.
2. Understanding the social and technological implications of information society.
3. Mastering basic concepts and skills necessary for using the new technologies.
4. Creating awareness of the potential and limits of the new technologies.

Underlying the design of the curriculum is the approach which draws on information science as the body of knowledge and skills required in order to generate, communicate and utilise information (Weiss, 1977).

Appendix 1 presents the major concepts, which were divided into four categories:

1. *Generation and organisation of information.* This category deals with the ways in which knowledge and information are created and codified (Cognition and perception;

symbol systems; writing and numbering systems; data collection; models; simulations; algorithms; etc.).

2. *Information transfer and distribution.* This describes the process by which information is communicated among the individuals comprising a given population (live populations, machines, or a combination of the two). Two sub-categories of classification are differentiated, according to the purpose of such transfer:

A. Communication: Here are concepts such as the formation of a communication circuit; means of communicating information; interpersonal communication and mass communication; man-machine communication, etc.

B. Storage and data processing: This includes concepts such as information storage, means for storing information; memory; information processing; the activity of the brain; the computer units, etc.

3. *Means and applications.* This refers to the identification of means (instruments, technologies) and the practical expression of information systems in social life, e.g. symbol systems, measuring and computing instruments; means of communication, communication technologies, decision-making processes, education technologies, art, etc.

4. *Effects and influences.* The social phenomena in which information systems play a major role, such as: the evolution of language, the print, electronic devices, education, mass communication, the management of complex social organisations; propaganda and advertising; leisure culture; social centralisation and de-centralisation; etc.

The contents have been arranged hierarchically, according to the age level, and a content-related focus has been determined for each level. The rationale for the planning and the detailed mapping appear elsewhere (Mioduser, 1982).

For further elucidation of the multiplicity of curricular, cognitive and organisational problems related to the development of this project the grade 6 level learning unit "Information in-Formation" was developed. This unit will be described in greater detail in the following sections.

II. "Information In-formation": a Learning Unit for Grade 6

This learning unit focuses on the various stages of information processing, as it takes place in everyday life, performed by the human brain, and with the aid of computers.

The objectives of the unit:

A. The student should know and understand the stages of information processing: input—processing—output.

B. The student should be able to describe daily phenomena in terms of the process: input—processing—output.

C. The student will experience processes of problem solving and decision making, and will be able to identify and describe these activities in terms of input—processing—output.

D. The student will learn how information is being processed by the human brain.

E. The student will understand that man has built machines, the speed and accuracy of which help in information processing.

F. The student will learn and understand the information processing by computers.

G. The student will develop a positive attitude towards modern technologies, through understanding the logic of the structure, mode of operation, capabilities and limitations of information processing systems.

I.   *The Principles Underlying the Development of the Learning Unit:*

Content Principles

A. *Technology as an extension of the natural capabilities of man.* Information processing technologies are perceived as an extension of the human nervous system, serving to increase man's capabilities in performing activities unique to him, namely, the activities of the human brain. The unit presents the information processing as it is carried out in the human brain and by machines built by man to help in performing these activities.

B.   *The influence of technology on the life of individual and society.* The various roles of information, and the changes introduced into the life of individuals and the society at large as a result of the development of information technologies, are illustrated by the unit through the presentation of a variety of examples for information processing and the application of computers in daily living.

C.   *The discipline and related skills.* The unit centres around the basic concepts of information processing and the technologies by which man is helped in processing information, as well as the intellectual and motor skills involved in applying this.

Curricular Principles

A.   *Acquisition of skills through experience.* The proposed activities involve experiencing the use of a variety of skills which the unit wishes to teach. These include intellectual skills (for example: selection among alternatives, problem solving; organisation and retrieval of information) and motor skills (beginning with model building and up to operating a computer).

B.   *Representation of the Immediate Environment of the Learner.* The topics of the activities are derived from the immediate, relevant, environment of the student: home, school, the apartment house, the city. There is also reference to information derived from newspapers, children's literature, activities at school, etc.

C.   *Variate instruction methods.* The contents of the unit are taught various means: texts for reading, games, building and assembling models, operating these models, keeping a diary, newspaper reviews, interaction with the computer, etc.

D.   *Self-teaching.* The unit enables the individual learner to advance independently. The presentation of the various topics, instructions for performing the activities, instructions and examples for keeping the diary and components of the models to be built—all comprise a part of the textbook.

2.   The Contents of the Unit

The Unit is comprised of four chapters:

A.   *The world of information.* This deals with identification of the roles of information in daily life: identification of the components of the physical environment which are related to dealing with information; the use of information by various role players; the role of information in decision making.

B.   *Information processing.* This chapter presents the basic concepts: input—processing—output; it also deals with identifying information processing in examples taken from daily life and discriminating between processes which people carry out in their brains and those performed through the aid of machines.

C.   *Information processing in the human brain.* This chapter focuses on the variety of information processing activities of the human brain, and the role of perception and memory in such processing. It also employs the concepts of input—processing—output in the analysis of examples drawn from everyday life.

D.   *The computer—an information processing machine.* This chapter introduces the structure and mode of operation of the computer, and the necessity of instructions by man for the operating of the computer; presenting the personal computer and its applications to daily life.

3. The Structure of the Unit

The unit is comprised of a textbook, containing various activities and reading excerpts, models that should be put, built and operated, games, computer software, enrichment notebooks and kits for assembling (for details see Appendix 2).

The four chapters all have a similar structure: Each opens with a paragraph presenting the core of the chapter and the related problems. Two to three major activities follow (for example: operating a model, press review, games). Throughout the learning the student is requested to compile a "diary"; following each activity in the chapter, he is expected to perform various tasks in the Report Notebook. At the end of each activity and each chapter reading paragraphs are given, which summarise and elaborate on the learned topic. Finally, there is the "Chippie report", which summarises the major concepts taught in the chapter.

Appendix 2 details the activities in each chapter, as well as the additional components of the learning unit.

4. The Development of the Unit

Two years were required to reach the present version of the unit, which includes models, games and computer software. The development has been accompanied by evaluation activities, including experiments in the classroom, carried out at various stages of the development, and consultations with experts in the various subject-matter areas, as well as in cognitive psychology and curriculum development. The present version is the final result of the above, and it will undergo systematic evaluation in the coming months.

An independent parallel project has been the development of the computerised simulation 'The Transparent Computer', that deals with the contents of Chapter Four of the unit. The development of this project has taken a year and a half, and the present edition is the result of systematic experimentation and evaluation. Detailed description of the rationale and the implementation of this simulation in the classroom appear elsewhere (Mioduser, Nachmias and Chen, 1984). The enrichment booklets, audiovisual materials and kits for assembling (items 6,7,8 of Appendix 2) are currently under preparation.

Appendix 1

MAN COPING WITHIN AN INFORMATION SOCIETY: Organisation of the proposed course

GRADE	1. DATA GENERATION, ORGANISATION, AND STORAGE	2. INFORMATION TRANSMISSION AND PROCESSING
1–2	Properties of matter. Classification-group theory and categorisation. Measurements-sources of information.	Basic communications circuit. Information transmission processes in the animal kingdom. Verbal and non-verbal description instruction and learning.
3	Signs and symbols (visual, acoustic, written, etc.)	Means of information transfer: voice, body, noise, instruments,

	Language (spoken, written).	writing, etc. Verbal and non-verbal communication. Communication patterns among animals.
4	Data organisation and classification: Coding-decoding, simple means of information storage, measurements, quantification, etc.	The input/processing/output circle.
5	Signs: conventions, supersigns, hierarchy, coding, bit chunking.	Transmitter/Means/Receiver/ Noise Signal: electric pulse, acoustic and optic wave, printed character, etc. Photography-Recording.
6	Organisation of simple information systems. The binary method. Advanced means of information storage.	Algorithm—Program. Senses and memory (LTM, STM). Computer components and their functions.
7	Probability information quantification. Frequency redundancy. Formation of supersigns (via complex or category formation). Sources and motivation initiating communication processes.	Techniques in the arts. Techniques in advertising and propaganda. Language, Education, Memory Role of communications processes in the formation and functioning of social organisations.
8	Flow chart-Probability Model- Simulation. Computer language. Feedback.	Man-Machine communication. Electrical circuit-Mechanical transmission. Analog and digital principles.

GRADE	3. INFORMATION SYSTEMS APPLICATION	4. OUTCOMES AND IMPLICATIONS
1–2	Various manifestations of the communication process in the environment. Simple measuring and calculating instruments. Trial and error method of problem solving. Carrying out instructions.	Perception of the environment: classification, organisation. Identification of functions in communication processes.

3	Symbol systems Simple Codes: Morse, flag signals, etc. Means of interpersonal communication. Coding and Decoding.	Transition to written language History of information transmission. Advent of printing.
4	Computation instruments: Simulation. Simple computation instruments.	Process optimisation by using auxiliary means for calculating and computing.
5	Mass-communications. Innovative ''media''. Communication occupations.	''Media'' and shape of society. ''The medium is the message'' Extensions of Man
6	Information systems in the immediate environment (classrooms, library, etc.)	Information explosion. Optimisation of the organisation and management of information systems.
7	Advertising and propaganda. Education. Arts. Social organisations (animals and man).	Changing relationships of individuals and society. Centralisation and decentralisation of social structures. Change of occupation profile. Education, Arts. Culture of Leisure. Shaping public opinion. Advertising, propaganda.
8	Information systems in the service of Man. Processes of decision making. Control Automatic regulation. The interplay between the individual and information systems.	Decision-making processes in bureaucratic and democratic organisations. Rearrangement of workforce (from manufacturing occupations to information handling). Leisure time.

GRADE 9

Project in one of the areas of the unit (e.g., information system for use in the classroom; ''animation film''; planning a publicity campaign using several media, etc.)

The following components will be part of this unit:
1 Concepts and processes such as: ''Intellectual Technologies'', Decision Making, Planning, etc.
2 Guidelines for project development:
2.1 Methodology: organisation, planning, examples, etc.

2.2 Techniques: use of instruments and application of various technologies.
2.3 Reference to sources: consultation, information, equipment, etc.

CONTENT FOCUS

Grades 1 and 2:	Basic principles of communications and automatic computation.
Grades 3, 5 and 7:	Communications and media.
Grades 4, 6 and 8:	Computers and information processing systems.

MAIN THEMES FOR EACH GRADE

Grade 1–2	Fundamentals.
Grade 3	Significance of communication.
Grade 4	Use of simple aids for data classification and organisation.
Grade 5	Mass communication.
Grade 6	Information processing.
Grade 7	Communication processes and the functioning of society.
Grade 8	Information systems (``Communication'') in a social context.

Appendix 2

1. The Subjects of the Learning Unit

Chapter 1: *The World of Information*

The roles of information.
Objects and instruments related to information.
People who use information.
The necessity of information.
Information in decision making.

Chapter 2: *Information Processing*

The process of input—processing—output.
The processing of information.
Information processing in everyday living.

Chapter 3: *Information Processing in the Brain*

The various roles of the brain in information processing.
Examples for information processing in the brain.

Chapter 4: *Information Processing by the Computer*

The computer as an extension of man's nervous system.
The computer units and their roles.
Program and programming language.
The personal computer and its application to daily life.

2.   The Contents and Activities of the Unit.

Chapter 1:   *Information on using Information*

(a)	Information, pigeons and the weather	: Introduction: the role of information today.
(b)	Excavation in a thousand years	: Studying our information age through an archaeological enterprise 1,000 years from now.
(c)	People use information	: A newspaper activity on uses of information in daily life.
(d)	The Mayor's decisions	: A decision-making cards game (see items 4 and 5).

Chapter 2:   *Information in-formation*

(a)	A storm in a cocoa cup	: Introduction.
(b)	Materials getting processed	: Learning input/output/processing concepts through building and working with a ''coining machine''.
(c)	The desperate detective and ''Identi-kit 572''	: Building and operating a model (see item 3).
(d)	Information getting processed	: Summary.

Chapter 3:   *Please Use Your Brains*

(a)	The Council of the Wise Men	: Introduction.
(b)	Please use your brains	: A series of activities comprising: association, identification, memory, computation, etc.
(c)	Our correspondent for brain affairs reports . . .	: text and discussion.
(d)	Info—An expert for information processing	: Game for building examples on information processing by the brain (see item 4).

Chapter 4:   *The Computer is No Wizard*

(a)	Man built machine	: Introduction.
(b)	Chippie 572—The forms-making machine	: (See item 3).
(c)	It's your turn to program Chippie 572	: ''Programming the forms-making machine'' (or ''The transparent computer'', a computerised simulation—see item 5).
(d)	A computer is no wizard	: Summary

3.   Models

On Chapter 2:	''Identi-kit 572'' (A machine for information processing). A model for composing faces on the basis of the binary principle.
On Chapter 4:	Chippie 572—The forms-compiling machine. A model of a computer and its operation. Includes the various

units of the computer. May be programmed to compile various forms, and demonstrate the stages of performing the program step-by-step, up to the final output stage.

4. Games

On Chapter 1:   *The Mayor's decisions:*
                        A decision-making game, involving decisions on issues concerned with daily life in town.

On Chapter 3:   *Info—An expert for Information Processing*
                        Puzzle of illustration of information processing in the brain. The students have to provide examples of effective and cognitive activities that lead from an observed input (a card he chooses) to the output.

5. Computer Software

On Chapter 1:   The Mayor's decisions (in planning):
                        A computerised decision-making game. In addition to the game included in the book, it includes:
                        — additional branching.
                        — various levels of clues and information purchase.
                        — graphic screens.
                        — rich interactive activity.

On Chapter 4:   "The Transparent Computer": A simulation of the computer's units and its mode of operation (developed in 1984):
                        A computerised simulation which may be programmed using a limited language. While running the program it is possible to follow the order of performance of the statements, the transition of statements and numbers among the different units, the changes occurring in the various units throughout the performance of the program and the various stages leading to the output.

6. Information Booklets (in planning)

On Chapter 1:   Articles and papers from newspapers on the issues of the development of the computer and information technologies, their application and uses in daily life, etc.

On Chapter 3:   Reading cards on the topics: The Brain, Memory Tree, Nervous System.

On Chapter 4:   The history of the computer, from the abacus up to the actual and prospected developments.

7. Audio-visual Aids

On Chap. 1–2:   The information Society.
                        A movie describing the applications of modern technologies to daily life. Composed of two parts:
                        A. The development of information technology: A short review, beginning with the alphabet and the abacy, and up to the computer.

B. The application of computers in daily living, and the prospects of future development.

8. Building Kit (in development)

On Chapter 4:   Build your own "processing machine":
Assembling models of the various units of the machine:
— binary adder.
— memory unit.
— screen or some visual output.
Including a manual with explanations, and the materials for assembling the models.

References

Abelson, P. H. and Hammond, A. (1977) The Electronics Revolution, *Science, Vol. 195, No. 4283.*

Bell, D. (1979) The Social Framework of the Information Society, in Dertouzos, M. L. and Moses (ed.): *The Computer Age—A Twenty Year View,* M.I.T. Press, Cambridge, MA.

Bolter, J. D. (1984) *Turing's Man, Western Culture in the Computer Age,* The University of North Carolina Press, Chapel Hill.

Chen, D. and Nachmias, R. (1982) The Use of Computers in Education (Hebrew), *The Computers in Education Research Lab., Research Report No. 1,* Tel-Aviv University, Israel.

Chen, D. and Novik, R. (1984) Scientific and Technological Education in an Information Society, *Science Education 68(4),* Wiley & Sons.

Dertouzos, M. L. and Moses, J. (eds.) (1979) *The Computer Age—A Twenty Year View,* M.I.T. Press, Cambridge, MA.

Forester, T. (ed.) (1980) *The Microelectronics Revolution,* Basil Blackwell Publisher, Oxford.

Klassen, D. (1981) Computer Literacy, *TOPICS—Computer Education for Elementary and Secondary Schools,* ACM Sigcse-Sigcue.

Licklider, J. C. R. (1979) Impact of Information Technology on Education in Science and Technology, in Deringer, D.; Molnar, A. (eds.), *Technology in Science Education—The Next Ten Years,* N.S.F., USA.

Luehrman, A. (1980) Computer Illiteracy, A National Crises and a Solution for It, *BYTE,* July.

Mcluhan, M. (1964) *Understanding Media: The Extensions of Man,* McGraw-Hill, New York.

Mcluhan, M. (1967) *The Medium is The Message,* Bantam Books.

Mioduser, D. (1982) *Man Coping Within an Information Society: A Proposal for an Instruction Unit,* MABAT Project, Tel-Aviv University.

Mioduser, D., Nachmias, R. and Chen, D. (1984) The Use of Computerized Simulation to Introduce the Functional Structure and Operation of the Computer, *Research Report No. 4.*

This course from Israel was developed as part of a wider programme concerned with environmental education. In the UK the motive for setting up a programme was rather different. It was largely devised because of the Government's plan to provide every school with a microcomputer—which, of course, means that every child in the country, from whatever background, is aware of the existence and capabilities of a microcomputer—though for many, it will so far have remained little more than a sophisticated means of playing games. The following paper by Trisha Strong presents the UK perspective on Information Technology in Schools.

INFORMATION TECHNOLOGY IN SCHOOLS—
A UK PERSPECTIVE
TRISHA STRONG

Why Have an IT Course?

In 1982 the British government began a scheme to raise public awareness about new technology. As a part of this scheme every primary and secondary school and tertiary college has been offered financial help to buy microcomputers. Now there is at least one micro in almost every school. At the same time the home micro market took off and now micros are common at home too. So new technology has made its presence felt in many aspects of life at home, at work and in the street. It is readily accepted by the young because to them it is a way of life, they have never known anything different.

It is we adults who find the rate of change so hard to handle, and schools are even slower to adapt to change than individuals. It is not surprising then to find that although pupils live in a high-tech world outside, in school the most advanced examples of technology are the video recorder (for which the teacher may have to gain a special certificate if s/he wishes to operate it), and the computers which have been bought because they were subsidised by the government.

Pupils live in a high-tech world, so how is education coping with the phenomenal rate of change in technology and the effect it is having on our society?

Structure of the Project

Computing is studied in many schools, by pupils of many ages, but mostly over the age of 14. However the examples of technology in daily use (video recorders, microwave ovens, telephones, teletext) seem very far removed from computers and how they work. Consequently any computing studies seem to have very little relevance if the aim is for the confident and efficient use of new technology.

This idea forms a major principle behind the Croydon IT Project. Information Technology is not about how computers work and how to program them, but gaining familiarity and confidence with new technology so that it can be used effectively and will help to cope with future change. The project aims to develop independent learners and informed consumers/users of new technology, capable of assessing their own information requirements and solving their own information problems.

A project was established by the London Borough of Croydon in September 1982 to produce a series of 12 pupil's books with supporting software and teachers' notes to provide the core for a course in IT. The course will be very practical though will not necessarily use a micro. However, trying to run an IT course without using a micro or any other technology would be like learning to cook by reading a recipe book and never making anything.

Each booklet will provide a complete unit of work on a topic in the syllabus. The series will not be sequential and, although one volume will be an introduction to the course and some will be aimed at the top end of the age range, the course can be ordered to suit the strengths and preferences of the teachers involved. Two booklets will be devoted to considering real-life applications and the resulting implications of IT. Through studying applications, the topics covered in other booklets can be put into context and the implications become meaningful.

There are different ways of presenting IT in schools. Information Technology can be taught as a separate course or it can be incorporated into each subject area as necessary. The latter model would be ideal as it would involve teaching with IT rather than about it. Unfortunately this model relies on the majority of teachers being confident and knowledgeable about the use of new technology and also on detailed communications between departments. Many schools and boroughs have opted for the former model which is less expensive on human resources and can be implemented immediately, recruiting one or two teachers from any department in the school.

In the Third World, the starting point is very different, pupils are less likely to be surrounded by examples of new technology at home, or in the street. However, when it comes to work pupils are likely to have the same technological needs as their European counterparts. Third World countries will not necessarily go through the same development pattern as the Western world, i.e. industrial revolution and twentieth century commercial development; they may well take a huge leap forward into today's technology, e.g. straight to satellite communications rather than via an extensive land line telephone network.

The whole gamut of technological development will exist within the possible experience of any pupil, so each of them must be prepared to handle the advanced technology they may come across at work later on and which may become a part of their daily experience sooner than we think.

What Goes Into an IT Course?

Information Technology has two words in the title, yet undue attention is always given to the second of those two. This emphasis on the technology side of IT serves to put off a great many teachers who claim to have no interest in, and no aptitude for things technical. The Croydon Project takes the view that IT is primarily about information and communication and about how the technology can be used to help us make the easiest and best use of the huge range of information which is now available for us. If this view is accepted then there remains no argument against incorporating IT into the school curriculum, since schools have always been about the acquisition, interpretation and presentation of information. It stands to reason also that the teacher becoming involved in IT need not necessarily be a technology expert, but should possess the communication skills and interest in technology to enjoy the course.

Looking at applications relevant to the pupil i.e. at home, school, work and leisure, will inevitably mean much of the material could also be encountered in other subjects, but the purpose here is to highlight the principles and to set the applications within the context of IT. Consequently the syllabus includes these areas:

What is information and how do we use it?
Information Skills.
How is information represented?
How is information stored and accessed?
How is information transmitted and communicated?
IT in education and the home.
The electronic office.
IT in monitoring and control.
The concept of a system.
IT in forecasting and simulation.
The applications and implications of IT.

Aims of the IT course

The general aims of the course are to ensure that all pupils:
(i) have an understanding of what characterises information processing— including control technology—and the use of computer-based information systems;

(ii)   are aware of major developments in technology affecting communication of information;
(iii) experience the use of modern communication devices and systems in some worthwhile way;
(iv)   use computers to perform useful activities which would be unduly tedious otherwise and hence develop a practical understanding of problem solving using computers;
(v)   have sufficient familiarity with common electronic communication systems to have no irrational fear of them;
(vi)   develop the understanding and confidence to use Information Technology to enrich their everyday lives;
(vii) develop a basic understanding of the principles underlying Information Technology so that they can cope with changes in technology.

IT and Learning

The microelectronic revolution has already had a great effect on the working and leisure activities of many people, but this is insignificant compared to the impact which will be felt when business machines, computers, telecommunication devices and domestic equipment become much more compatible and interconnected. Today's children must be prepared to understand the causes and effects of these almost inevitable changes.

These features will make new demands on education. Employment patterns will be in the knowledge industries. Society needs to transfer labour from the manufacturing industries to knowledge and information-based industries.

Education is more than just compulsory schooling, so the schooling which is given must provide young people with the ability to learn for themselves. Once the school environment is left behind then the learner has to depend on self-discipline and information/study skills to succeed. The styles of learning exemplified by the Open University, the proposed Open Tech and the various Distance Learning and Flexi-Study methods now being employed are likely to become more common, and it will be essential for all pupils to acquire sufficient confidence and skills to use these systems.

There are various initiatives to continue the UK's involvement at the forefront of the IT industries eg. The Alvey Commission. Such initiatives recommend the urgent provision of trainers for both industry and post/undergraduate studies, This is a short term solution with an output in the next 3–5 years, but investment is needed for the long-term future. By setting up suitable courses at school level now, we are making a start on providing a development pattern for future human needs.

The Computer in School

The computer in school is seen as a tool to enhance classroom learning. This can be achieved in many ways, through the computer organising the learning as with CAL and CML or through the style of activity the computer is used for. Computer programs can be used to develop a set of skills and abilities which are developed across the curriculum, for example: decision-making skills, planning and organising skills, strategy determining skills, and problem solving.

Games and Simulation

The excitement which children derive from video games can and has been built into games designed for educational purposes. When the computer is used in this way by groups of children, purposeful discussion usually takes place about a common problem.

Discussion like this will clarify thinking and provide opportunity for social interaction with the need to listen to other people's points of view. The computer frees the teacher to be an adviser, to assist and discuss in a creative and dynamic sense.

Problem Solving

With a problem solving program the children can set themselves problems which are real to them, and therefore they have the necessary motivation to solve them. The children become totally involved and can take the initiative in directing themselves through. They can formulate their ideas, test them, correct their mistakes, reformulate, retest and in so doing devise their own system and develop their own strategies. Children learn by their mistakes and often more than one route to the solution can be identified.

This use of the computer gives children opportunities for co-operation, sharing ideas, putting forward arguments and decision making. They learn to analyse their mistakes and use their powers of logical thinking. They need to be able to think ahead and they require long periods of concentration. This way of using the microcomputer in school is probably one of the most interesting and demanding and well worth investigating.

Information Handling

Programs which introduce children to information handling techniques can be a powerful tool for independent learning in the classroom by encouraging careful thought about data collection and selective use of reference sources. Children are introduced to the skills of classification which leads to an understanding of the logical operations to obtain information. When children have created a file of information, they can then concentrate on drawing conclusions by asking questions to sift and sort out particular records that meet specific criteria.

By learning how to handle, store and retrieve information which they have collected from their immediate environment by making a survey or using reference books, the children are learning about the nature of that information.

Equal Opportunities and IT

This project is determined that all children should be provided with the education to enable them to cope with the technology that will become an integral part of their way of life. To avoid the risk of IT becoming another male-dominated reserve as computing, maths and physics are in western Europe and the USA, it is essential to involve everyone. The effects of the technology are not confined to men only and girls must not be discouraged from enjoying and benefiting from the course.

It follows that many teachers will need to improve their own attitudes to technology and, in particular, to ensure that sexist attitudes which allow women to use office products and men to use obviously mechanical or electronic devices, are not transmitted to the next generation.

Never in the development of the education system has an area of the curriculum been subject to such commercial pressure and influence. Unlike mathematics and physics, children are more likely to develop their interest in information technology at home than they are at school—at least boys are more likely to have a computer in their home than girls. These figures are hardly surprising when it is apparent that many manufacturers of both computer hardware and software are consciously directing their advertising at the male consumer, whatever his age.

It is important therefore that schools consider carefully how best they may be able to counteract the influence of external factors on the IT education of their children.

IT Education

If, as has already been stated, the IT Project is about communication and information, then pupils will learn to select the appropriate technology for their information needs. This need not necessarily include the computer. An appropriate information transfer technique may be a storyboard, or an information source may be a collection of newspaper clippings.

To equip the future generations with the skills envisaged for their information based society, pupils need to know how to read, to listen and to discuss. They must learn how to present information effectively and how to assess the relevance and accuracy of information. Most importantly they need to question the validity of innovation, to discuss the advantages and limitations of new technologies and to consider the implications for their own future needs.

Points to Remember

1. Information Technology must be a practical activity which builds pupils' confidence to use computers and other information sources.
2. Knowing about computers means knowing when and how to use them rather than knowing the details of how they work.
3. Pupils should learn to use computers sensibly, i.e. by using the appropriate software, to perform tasks which would otherwise be unduly tedious.
4. It is likely that sensible activities using computers have a positive benefit on girls' attitudes if started in primary school.
5. Information Technology can be taught either as a separate subject or as a co-ordinated part of many other subjects.
6. Teachers need training over a reasonable period of time to become confident and proficient in the use of computers.
7. Teachers can help pupils appreciate IT by using appropriate means of communication in their own lessons.

The Information Technology Project is published by:
Addison-Wesley Publishers Ltd.
Finchampstead Road, Wokingham, Berkshire RG11 2NZ, England

One of the aims of this book is to help countries to benefit from the experiences of others in educational development. But, of course, in order to appreciate the various features of a programme it is necessary to know something of the background against which it has developed. The following paper by Zsusan Szentgyörgyi describes the background and the progress that has been made in teaching informatics in Hungary.

INFORMATICS—ITS USE AND EFFECTS IN HUNGARY
DR. ZS. SZENTGYÖRGYI

Hungary is situated in Europe's middle, in the Danube basin on an area of 93031 km². Her population in 1981 was 10,7 million with an average density of 115 persons/km². (The density of the world's population is 31, in Europe 64 persons/km²—data from 1978.) Regarding the territorial distribution, 19,3 % of population live in the capital, Budapest (2,06 million) which is at the same time the administrative and industrial centre of the country. In other five cities (Debrecen, Györ, Miskolc, Pécs and Szeged) live about 8 %

of the population. According to the GDP, Hungary occupies the 32.–33. place among the 165 countries of the world and the 15.–16. in Europe.

In 1984 the number of computers (mainframes) was about 1000, while that of mini- and microcomputers was between 5000–6000—and is evenly growing. An important fact is the changing structure of computer usage: whereas in the first periods of the computerisation (i.e., in the late sixties and the seventies) computers could be found mainly in centralised applications (databanks, budgetary, state planning modelling etc.), and in some research work, just now—grasping the possibilities of the new, cheap, easy to learn and handle devices, such as personal computers (PC), local networks—a new flux can be observed: the usage of computers in every branch of economy; in industry, agriculture, transport, human services, mass media, and even at home. Computers are used at workplaces in construction and machinery design (CAD = Computer Aided Design), in metal cutting and processing industry (CNC = Computerised Numerically Controlled machine tools and systems), in light industry for optimising the quantity of material to be tailored, in chemistry for process control, and in agriculture. A new process is spreading in the computerisation of administrative work, the so-called office automation (text-processing, data files, electronic mail). About 2 years ago a new—at present only experimental—service was introduced in Hungarian TV-broadcasting: the teletext information.

While technology moves miles ahead, there remains a vital question: how can people acquire the knowledge to use, and—as much as possible—enjoy its benefits? In Hungary workers do not have to face unemployment; all the people—inclusively women—can find a job in their active life; the Constitution guarantees work for everybody. Nevertheless, there are facts which are to be considered in the forthcoming years:

   (i) the government in Hungary exerts great effort to raise the productivity of work which is rather low compared with other countries in Europe;

   (ii) the employment structure shows a highly transitive facet during the last 2 decades (see Table 1);

   (iii) by the introduction of new technologies (especially microelectronics, informatics, robotics) new jobs have been created, while some traditional ones are submitted to change or even full extinction.

TABLE 1

THE DISTRIBUTION OF EMPLOYEES IN THEIR ACTIVE AGE
IN THE DIFFERENT BRANCHES OF ECONOMY

	1971		1981		(1981)	
	10^6	%	10^6	%	M	F
Industry & building	2,16	43,1	2,05	40,9	1,24	0,81
Agriculture	1,23	24,6	1,03	20,5	0,62	0,41
Material services	0,85	17,0	0,96	19,2	0,53	0,43
Non-material services	0,77	15,3	0,97	19,4	0,37	0,60
	5,01	100,0	5,01	100,0	2,76	2,25

These components are, of course, very strongly interrelated since the main source in increasing the productivity level is just the introduction of new technologies, while, on the other hand, job-conversion and job-erosion caused by this process imply fundamental changes in the employment structure.

Future-telling oracles say that informatics will have a revolutionary impact on the coming age of the twenty-first century. And really, the truth of this statement can be proved with great confidence regarding the trends in the last and present decades, and extrapolating the expectable processes and effects both in technology itself and its societal consequences. The way a society can give a proper response to the challenge of such a rapid and deeply overthrowing process we witness in informatics, greatly depends on the strength, traditions, elasticity and fundament of its education system.

There is a multilevel education system in Hungary. Compulsory schooling begins at the age of 6. Kindergarten between 3 and 6 prepares children for school (for those who have not attended kindergarten, preparatory courses are organised). In school-year 1980/81 66 % of the children went to kindergarten. Primary school compulsory for everyone under 16—consists of 8 terms. In 1980 95,2 % of schoolchildren finished all the 8 terms—including those at evening- and correspondence courses, as well as at institutions for handicapped children.

Further educational possibilities after primary school:
  (i)   vocational training.
  (ii) secondary (grammar) school.
  (iii) middle-level training colleges.

Vocational training generally lasts for 3 years. Special emphasis is laid on practical education and those subjects which belong to general culture. In 1981 268 schools of this kind were operated in Hungary for 31 professional groups.

Secondary (grammar) schools prepare children for university. An interesting example proves that girls consider these schools a possibility to postpone final decision in choosing profession: in 1981 the ratio of girls at secondary schools was 57,6 % (while in vocational training 31,6 % only). Middle-level training colleges combine general education with learning of a profession.

Further education is carried out on two levels. Colleges provide first of all practical education during 3 years (work engineers, primary schoolteachers). At universities theoretical education is emphasised (5–6 years). At all education levels there are evening- and correspondence courses, as well. Table 2 shows the distribution of students in school year 1980/81 at day-, evening- and correspondence courses together.

TABLE 2

THE DISTRIBUTION OF STUDENTS IN SCHOOL-YEAR 1980/81

Education level	10^3 persons	%
Kindergarten	478,1	20,9
Primary school	1177,8	51,4
Secondary school	342,2	14,9
Vocational training	154,1	6,8
Further education	101,2	4,4
Institutions for handicapped children	37,3	1,6
	2290,7	100,0

In addition to schools there are various professional training- and postgraduate courses. In 1979 223 thousand persons participated in manual workers' courses (e.g. health, commercial, middle-level technological) and 111 thousand in postgraduated courses. Postgraduated and training courses teaching new professions serve the obtaining of up-to-date knowledge deriving from the introduction of novel technologies (technological innovations). Traditional school can follow these with considerable delay only, while the courses apply quick, flexible methods. A good example for this is the education of computer technology. Computer technology and programming have been taught in some secondary schools—especially in middle-level training colleges—already for 5–8 years. From the school-year 1983/84 on every secondary school has had personal computers for direct use by the pupil. In addition, every college and university provide basic education in computer technology (not mentioning mathematical and technical faculties teaching professional knowledge).

The concept of the development plans for education and training informatics, computing and microelectronics[3] in Hungary contains three general requirements which are to be achieved in the next years:

(i) due to the fundamental effects of informatics on the whole society, a common programme has to be worked out for the entire cross-section of the education system taking into consideration all the requirements and interrelations of the traditional subjects with informatics, eliminating parallelism, and creating new subjects if it is necessary;

(ii) the synthetic feature of informatics is to be used for integrating the knowledge given by the different subjects into an overall system;

(iii) the education system of informatics has to form a hierarchical structure in which knowledge given on higher levels should be based on those received on the lower ones.

Of course, the members of the society do not participate evenly in the education of informatics—it depends on their age, abilities, job requirements, and their interest. In a primary approach we can state that by the end of the century:

—rudimentary, basic knowledge on informatics have to be acquired practically by everybody (similarly, as literacy is a basic requirement to all members of an advanced society);

—people who need informatics at their work are to be able to use it effectively;

—experts of different levels in informatics will be educated and trained in an appropriate number (skilled workers, technicians, teachers, engineers etc.).

In order to accomplish this rather ambitious conception, financial and mental resources should be made available (hardware devices and systems, networks, software means, skilled personnel, books, training courses etc.). According to some preliminary recommendations for the coming 5-year plan all primary schools in the country have to have 3–5 simple microcomputers (so-called school-computers (SC)), in each of the nearly 750 secondary schools there should be about 20 SCs, and at the universities computer laboratories are to be established on basis of professional personal computers (PPC)—about 8–15 in each laboratory. In addition to the introduction of SCs and PPCs at schools and universities, the establishment of regional education computer centres (which started in the seventies) is to be continued, and they are to be integrated into local and regional networks.

Another significant forum for professional courses and lectures is provided by the Federation of Technical and Scientific Societies (MTESZ) and its member-organisations which have already shared an extremely useful part in rendering knowledge on new technologies for engineers and other specialists (not only in informatics and related themes but in biotechnology, physics, agriculture etc. too).

Finally, I intended to give you a short account on the Hungarian women's participation in new technologies. I am afraid, however, that I cannot give you any specific reference to informatics. The share of women in these fields' jobs is quite high on the lowest or lower levels (e.g. operators, junior programmers), and is decreasing stepping higher on the hierarchy—which is a phenomenon corresponding to the overall job-distribution in the country. It does not mean, however, that women are banned from higher jobs. They can learn and be trained with men but women do not get jobs of higher ranks.

—partly due to the lack of aspiration,

—partly because of the household and child-care burdens, and

—in quite a significant part in consequence of old traditions and prejudices.

On the other hand, the trends are encouraging because among young people these factors are less expressive and determinant as among the older generations. And I am convinced that it is the new technologies which bring about new facilities and possibilities for women, such as e.g. home terminals for young mothers who want to continue working while staying at home with their babies, new forms of jobs etc.

Notes

1. Computing and informatics are often used inaccurately as synonyms. They are, however, different notions: computing stands for the means—i.e. equipment, systems, software, while informatics is their application.
2. Friedrichs, G. and Schaff, A. (1982) *Microelectronics and Society. (For Better or for Worse.)* Pergamon Press.
3. In the following we refer to them only as "informatics".
4. Hungary's economic development is planned in 5-year long periods. The next will begin in 1986.

Microcomputers are much more widespread in Zimbabwe than in most countries of the Third World and a combined project of the Ministry of Education and the University has already produced a course on Information Technology under the title "Zinformatics" The following paper by Mike Robson and Peter Towse "Zinformatics: An Information Technology Teaching Project in Zimbabwe" gives details of the course.

Zinformatics

Open your newspaper here in Zimbabwe any day and you will almost certainly find a report on some new development in the field of information. It might be a report on a new telephone link to other countries, the introduction of a new computer link between branches of a building society, or the introduction of semi-automatic equipment in a local engineering company. These are all examples of the new information technology, which is already a reality here in Zimbabwe.

Since information technology is already part of the Zimbabwean scene, perhaps it is time that Zimbabwean children began to hear something about it at school. This is where the Zinformatics Project may have some part to play.

ZINFORMATICS is a joint project of the University of Zimbabwe and the Ministry of Education, and aims to explore new ways of teaching about information technology in Zimbabwe's secondary schools.

Information technology (IT) is a combination of information theory, computing science, and electronics. It is concerned with the techniques used for the storage, retrieval, transmission and transformation of information.

IT is not really a new subject. For as long as language has been with us, information preserved in the memories of old people has been passed on to younger ones, so that information was in effect "stored" in a living medium. And when one human group invaded, traded with, conquered, or otherwise interacted with another human group, information was transmitted and transformed. Thus storing, retrieving, transmitting and processing information has been part of our way of life since time began. However, with recent dramatic developments in electronics and computing science, suddenly information can now be stored and accessed, transmitted and transformed in enormous quantities, and at fantastic speeds. It is these developments which are changing the world about us at this very minute.

Already, of course, several schools have purchased computers. Too often the teachers are not really sure what to DO with the computer when they get it. The Zinformatics Project aims to help teachers to make better use of these expensive resources, and to show just how much meaningful teaching about information technology can be accomplished with the barest minimum of imported equipment.

The project tries to do this by producing books, and kits of tools and materials, for use in trial schools. Although cheap computers are supplied as part of the Zinformatics Kit, children using the Zinformatics materials are encouraged to make much of the equipment themselves.

Computers are much more widespread in Zimbabwe than in other Third World countries and it is important that, on leaving school, pupils should understand the important role that these machines play. It is more important that they understand what computers are, what they can do to make our lives better and how to handle them than that they are able to understand the higher technology of these machines and how to program them. For this reason, computers are introduced against a much wider background of information. Thus, the course is all about information, not merely about computers. In the West the word "informatics" has been coined to mean for most people "the study of information in all its various aspects" and hence it is inevitable that a project introduced in Zimbabwe should be christened "Zinformatics".

The work covered is divided into nine units or modules:

1. The nature of information
2. Information storage and retrieval
3. Transmission of information
4. Information processing
5. Information in the office environment
6. Control systems
7. Simulation and modelling
8. Information skills
9. The information revolution and the future.

There will be four books for each of these modules. One is a book of challenges which sets the pupils tasks. A second is a teacher's guide, which provides the teacher with background information and makes it easier for him to ensure that the tasks can be carried out successfully. A third is a book of reading materials for the pupils to help them see the tasks they have carried out against a wider background; this material is orientated towards a Zimbabwean, or at least African, setting. For instance, there is an article on talking drums, a unique and distinctly African way of transmitting information which, for its time, was highly sophisticated and comparable to any form of information transmission then practised. Finally, there is a book of puzzles and problems, which will extend what the pupils have learned and lead them on to further exciting and interesting things to do.

Schools which are not part of the Zinformatics trials will be able to purchase copies of the books as these become available. Photocopying will be freely permitted, provided this is not done for resale.

It is not possible for the work to be introduced as part of the normal school timetable and so it will be covered during the time devoted to club activities in the afternoons, for a total of about four hours a week. In this way it is not vying for a place on an already crowded timetable, but it is being treated with great seriousness and it is hoped to introduce a properly validated certificate at the end of the course for pupils who have shown a satisfactory standard of achievement.

The material is being produced by the project's directors, Mike Robson of the University's Computing Science Department and Peter Towse of the Department of

Curriculum Studies. The project will encourage teachers to think of the many ways in which they could use the computer as a teaching aid. Thus, it will be feasible to expect in time the production of local software, which will be more appropriate to our Zimbabwean setting than imported software—and it will save valuable foreign exchange. Eventually, it is hoped that some elements of the project might be incorporated into different subject syllabuses. For example, work on libraries could find its way into the English syllabus; the use of satellites in geological prospecting and weather forecasting could move into the geography syllabus; and control systems could merge into the science syllabus.

The Project is operating on a very small budget, without any aid from International organisations. However, the University's Research Board, together with generous assistance from eleven local companies lead by Turnall Holdings, has made it possible for at least a start to be made.

Essentially, these firms have donated sums of money to introduce kits into a small number of trial schools selected by the Ministry of Education. A kit includes not only a Spectrum computer with monitor and cassette recorder but an interface box which enables the computer to respond directly to information received from outside, together with a collection of simple devices such as morse keys. Fig. 10.1.

FIG. 10.1. Morse code communication equipment can be easily fashioned from scrap material.

The project is presently being tried out at three schools in very different settings, at Kawondera Secondary School in Zvimba and at Mount Pleasant School and St. Peter's Kubatana in Harare. The experience at Kawondera is particularly interesting for, although the school lies close to the Chegutu to Chinoyi road, it is a typical rural school without electricity and the computer is operated from a 12 volt lorry battery! The pupils see nothing unusual in this and are progressing every bit as enthusiastically and successfully as their urban counterparts. As someone at the school remarked, "computers are more concrete than the periodic table".

UNIVERSITY OF ZIMBABWE
Department of Curriculum Studies
ZINFORMATICS PROJECT
Publication Programme: Draft outline

There will be NINE modules in the Course, namely:

MODULE A   Nature of Information.
Information representation in analogue and in digital form; the consequences for accuracy in choosing one of these; the measurement of the quantity of information—bits and bytes; coding systems.

MODULE B   Information Storage and Retrieval.
Early forms of information storage in Africa.
Measurement of quantity of information stored.
Alphabetical ordering: various examples, including telephone directory, yellow pages, library catalogue and filing system, dictionary.
Binary decision search using hand-cut cards.
Storage on magnetic tape, storage in RAM and ROM, elementary sort routines.
Use of elementary filing software on computer, e.g. VU-FILE.

MODULE C   Transmission of information.
Early forms of information transmission in Africa.
Measurement of the quantity of information transferred, and measurement of the speed of transfer (bytes; baud).
Various media for carrying signals:
—light: semaphore, morse heliograph.
—electricity: morse telegraphy, telephony, the telephone exchange.
—radio waves: morse radio-telegraphy with spark transmitter, radio-telephony, earth satellites and world-wide telecommunication systems.
Sending pictures:
—serial transfer of pixels by voice.
—serial transfer of pixels by morse.
—parallel transfer of pixels by optical fibres.
—serial transfer of pixels by opto-electronic scanning and computer imaging.
—television.
—landsat.

MODULE D   Information processing.
Case Study: the 1983 Zimbabwe Census.
Statistical procedures as techniques for processing information for presentation in readily assimilable form.
Files, and the use of a computer in merging, sorting, extracting and collating information.
Graphical and other methods of presenting the results of data processing.

MODULE E   Information in the Office environment.
Office filing systems.
Accounting procedures.
Use of elementary accounting and forecasting software on computer, e.g. VU–CALC
(spreadsheet methods).
Word-processing.

MODULE F   Control systems.
Analogue to digital, and digital to analogue conversion.
Study of the idea of feedback through the exploration of various devices, including:
—the flush toilet.
—self-regulating borehole pump/reservoir system.
—the thermostat.
Study of the ideas of digital control, sensors, feedback, and servo-mechanisms
through the exploration of various devices interfaced with a computer, including:
—speed governor for a motor.
—automatic weather-data collection and recording.
—linked traffic lights.

MODULE G   Simulation and Modelling.
The business game: use of a computer to simulate the problems and processes
involved in running a small business.

MODULE H   Information skills.
Learning as a life-long process: how to learn:
—distance learning methods.
—how to use a library.
—how to find the right books and how to use them.
—other learning media and how to use them.
Analysing, processing, synthesising, recording and presenting information.

MODULE I   The Information Revolution and the Future.
Social implications of the new technology for the Third World.

The course is a very practical one and offers opportunities for "hands-on"
experience of hardware applications as well as for more theoretical studies of
the topic. "Toys through a time-warp" explores the philosophy of the course in
more detail.

Toys Through a Time-warp

In a well-known science fiction story, children's toys from the distant future pass
through a time-warp or some such nonsense, and fall down—when into our present
time. By chance, these toys are found by children; ordinary children, such as yours or
mine. They play with these toys from the future. But these are no ordinary toys; they are
toys designed by adults for teaching their children certain very advanced ideas about
space and time. Through playing with these toys, the children's minds become shaped
and transformed by them. At the end of the story the children disappear, moving off into
another dimension which the toys have revealed to them, leaving behind their helpless
parents who are quite unable to comprehend what has happened.

There is a vitally important idea here: toys can transmit major mind-transforming concepts, and transmit them more effectively than the written word; more effectively even than that most revered of all educational institutions, the elderly man with a piece of chalk.

Nothing new here, of course. Remember your Meccano set? (American readers please read that as "Erector"; others younger than me please insert "Lego".) Is there any engineer at work today whose world-view was not shaped by the order and constructed logic inherent in a simple Meccano set?

For children in the First World, of course, Meccano is dead. The modern equivalent of Meccano is the cheap microcomputer. When Clive Sinclair gave Britain the ZX80,81 and Spectrum, he gave British children toys through a time-warp. These toys cost about as much as a good dinner for two at a decent restaurant. Through one of these toys, computer engineers talk directly to children they have never seen.

Probably, Sir Clive felt few pedagogical imperatives when he designed these toys. But the next toy to come along was the BBC micro, which was certainly designed and intended specifically as a toy for learning. A whole generation of British children is now being shaped and transformed by the advanced concepts so brilliantly built into the "Beeb" and the Spectrum. These toys shout and scream ideas quite silently at the cleverer children, while many adults are completely unaware of what is going on.

But what of the Third World? Yesterday we were in Zimbabwe, at our University's Open Day. Our doors thrown open to all, we were visited by about 16,000 children, and a substantial proportion of these children found their way to the Hardware Laboratory of the Department of Computing Science. It is difficult to convey to you some feel for what was happening there. Thirty young first-year students in the room had projects on display. You must picture crowds of youngsters ten-deep about each one, pressing close to catch their words.

For example, one student had a working model of a lift (USA read elevator). This consisted of a very long glass tube made from an old fluorescent lamp. Above the tube, near the ceiling, the student had mounted a 20p electric motor, which by means of elastic bands and pieces of knotted string moved a magnet, disguised as a "lift", up and down inside the tube. Stuck with sellotape to the outside of the "lift shaft" at regular intervals were some 30p reed relays, which detected the arrival of the "lift". This simple model was entirely controlled by a Spectrum computer, with the DCP interface attached.

For two whole days the student talked himself hoarse, answering a rain of questions about control technology and computer programming from an absorbed and excited audience. This audience had never seen a computer before, but seemed quite easily to grasp what was going on. And the student himself had only met this particular piece of equipment, and the BASIC language, four days earlier. (BASIC is normally banned in our Department.) Certainly I taught him nothing; he learnt it all from these toys.

Nothing very strange about all this, except that, after all, it was in Africa.

The rest of the room, filled with Spectrums and Beebs, presented a similar picture. One student was printing out digitised photographs of visitors for a patient queue which stretched beyond the door. Another student had fitted electric contacts to the tie-rods of a kiddy-kar, and had connected these through a web of wires to the keyboard of his Spectrum loaded with "Chequered Flag", so that a child seated in the car could steer it and accelerate it on the colour monitor around the racing circuits of the world, to the sound of screaming tyres and roaring engine.

Significantly, very few adults were aware of this amazing happening. Officials from the Ministry of Education did not visit the scene. They found their way instead to the prize-winning display in the Education Department, where they gazed with considerable satisfaction at some excellent examples of wooden handicrafts and dyed fabrics borrowed from a local training college.

It seems to us, therefore, that the educational potential of toys has been missed by many educators. From Unesco in Paris we receive most excellent books on Science Teaching. We would receive from the same source most excellent posters such as Man and the Environment, if only we could afford them. But we don't receive any toys. Perhaps there is a Department concerned with designing toys for the Third World; toys with advanced concepts built into them; toys to cut out the teacher as middle-man and make direct contact with minds and hands; perhaps there is such a Department but we haven't found it.

Surely, if toys are really such potent and powerful instruments of self-education, we should take them seriously. They may well be subversive. There is no reason to believe that toys designed for the First World are suitable or even desirable in a developing country. Certainly, such toys are horrendously expensive for us. What costs in England no more than a decent dinner for two, costs us in Zimbabwe the equivalent of a year's wages for a well-paid worker in the construction industry. Could not we in the Third World develop our own toys, better suited to our needs, and (through reduced imported content), cheaper?

FIG. 10.2. A toy designed for first year university students reading computer hardware.

At the Department of Computing Science of the University of Zimbabwe we are making a few stumbling steps in this direction, in the form of the Zinformatics Kits.

Nothing very new of exciting here. Just a cheap cardboard suitcase with wooden frame inside supporting a hinged inner work surface made of "pegboard" Fig. 10.2. The inside lid of the suitcase contains a pocket for notes, diagrams and instruction cards, or for a set of simple hand tools. The body of the suitcase, below the work surface, holds a power transformer and rectifier unit; the rest of the space is taken up by a motley collection of hardware items. These items can be bolted on to the work surface in various configurations.

For example, one of our more advanced kits is intended for use, and is being tested by, university students in the first-year course in computer hardware. Students build the entire kit for themselves, and use it on the course.

First they cut, drill, hinge and glue the framework together, which is a novel experience for many, particularly the girls.

Next they build the power supply. They draw the circuit (using matchsticks and a locally produced protective paint) on copper clad circuit board material, and drill the holes. Then they etch the boards with a peroxide/acid solution to remove unwanted copper. The few cheap components needed are then soldered onto the board, and the board glued on to a suitable transformer. This then forms the power supply which is fitted inside the case.

Next the students build the hardware items needed. These include a number of holders for 74 series chips, plus logic switches, a logic probe and a clocking device. The logic probe is essentially a 7406 plus two LED's in a plastic tube; the clocking device is a 7400 flip-flop, flipped by twanging a piece of curtain wire. The foreign-currency component of these items comes to less than a pound.

With these and similar items students perform many of the traditional experiments with logic gates, flip-flops, shift registers, RAM, and so on. But because the whole thing is so cheap, they are encouraged to take the kit home and do the experiments in their own time.

After all, it is only a toy.

As an example of a kit for use at a lower level, there are those designed for schoolchildren. A class in each of our trial schools is provided with a Spectrum, an I-pack interface, a monitor and a tape recorder. These precious and expensive imported items are carefully looked after. But in addition, children are supplied with a Zinformatics Kit containing bolt-on items such as relays, motors, switches, lamps and so on, together with connecting wires and simple tools. With these they assemble interfacing projects very cheaply, and bring them back to school to try them out on the sparse computer equipment.

In one very simple project, children built a Morse key, and on connecting it to the computer were able to make the computer print out letters in response to Morse code signals from the key.

Results of our initial trials with such toys have been encouraging, but we cannot yet report any dispassionately observed, adequately controlled research findings. However, we are convinced that as prices fall in respect of some of the flotsam and jetsam of the Information Technology revolution, there will continue to be interesting and exciting work to be done in designing toys to bridge the time-warp between the developed and the developing world.

Appendix 1
Case Studies from Various Regions

1
Transfer of Biotechnical Information for Agricultural Development

ABRAHAM BLUM

Department of Agricultural Education,
Hebrew University of Jerusalem, Israel

A Model and Case Study of the Transformation Process

The Problem

Over 500 million people are undernourished or suffer from malnutrition. Partly because they are poor and have no land on which to grow food, even on a subsistence level. Perhaps to no lesser degree undernourishment and malnutrition are connected with a lack of knowledge, how to grow more food and how to improve the nutritional balance in their diet.

Agricultural extension techniques like the Training and Visit approach (Benor *et al.*, 1934) have shown that dramatic improvements in the agricultural output by small farmers can be achieved by introducing no-cost or low-cost practices like planting in rows with optimal spacing and depth control, correct weeding and so on. The problem is that the basic scientific knowledge is around, in most developing countries, but it does not reach the rural-agricultural community.

In this paper I shall present a tentative model which describes six major factors which modify the process of applying biological knowledge to rural development (Blum, 1982). I shall then illustrate the working of the model in a case study and repeat the truism that each case is different.

165

The Model

Biological knowledge, as such, has no purpose. Like other scientific knowledge, it is created and accrues in the process of wanting to know more. Usually, biological knowledge is sought without intention to solve practical problems, but sometimes the researcher points out possible future applications. A classical case was von Frisch's discovery of the bee dance patterns. At the end of his original paper, von Frisch (1946) stated:

> To introduce the procedure into practice would be a gratifying task for bee experimental stations. Then, as often before, the fruits of theoretical work will serve practice. As much as I hope and expect this to happen, allow me the confession: The purest joy remains that of insight, free of the earthly touch of a useful exploitation.

In other cases, basic biological research is done on a research grant which specifies that the knowledge sought should be relevant to a given bio-related problem. Thus, much of plant growth research was conducted in laboratories of agricultural stations, and much of bio-medical research is done in hospitals. Probably in most cases biological research is conducted without any practical implication in mind. Only when biological knowledge is reviewed and researched for those elements which have a direct bearing on possible community development, specific and relevant elements are identified. In this first step, the vast amount of biological knowledge is sieved through a grid of data on the community to be served and on its development problems (Fig. A1).

Actually, in most cases in which biological knowledge is applied to solve problems in a community, the action is triggered by researchers or intermediary change agents who identify the relevant biological knowledge which can be used to solve the problem. This happens in the stage which in the model is called Applied Bio-based Research and Development (R&D). This R&D work is done sometimes in a biological research institute, sometimes in an agricultural, nutritional, medical or other R&D institution, and sometimes even closer to the community to be served—in a regional extension or professional-educational centre. Only that knowledge which can be applied to solve the problem or can suggest an approach to solve it is considered. It is further refined and developed into recommendations on how to solve the biological aspects of the community problem involved.

Before these recommendations can be turned into plans for action, they must be tested for three constraints: available technology, economical feasibility and communication means. If the bio-based knowledge cannot be applied because of lacking technologies, or if it can be applied, but the technology is not economically feasible, or even if the solution is both technologically and economically acceptable, but does not reach the decision-makers in the community in a way understandable to them, no action towards community development can be expected. Such actions will only occur, if and when

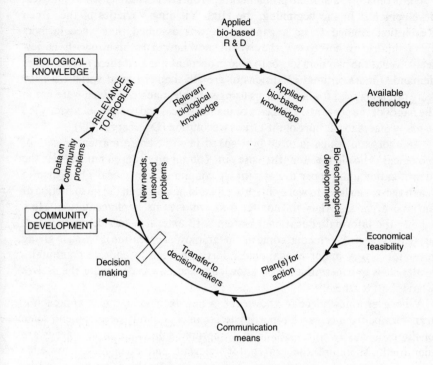

FIG. A1. The "biotechnical knowledge for rural development" cycle.

decisions in the right directions are taken and implemented at the community level.

The model should also remind us that we cannot hope for biological knowledge to affect community development by a simple diffusion process. If biologists want to reach the community and help to solve its problems, they have to co-operate with technologists, economists, communicators, educators and decision-makers.

Biotechnical development can be based on biological technology, e.g. in plant breeding, or it can be technical in the common sense of the word. In both cases the technology must be appropriate to the technical and economic capacity of the community to be served. This could be expected to be obvious, but it is often overlooked in practice. Many of the technological gadgets donated by rich nations to developing communities became white elephants, when no maintenance facilities were available and spare parts were unavailable or too expensive.

Unfortunately, a similar problem of appropriateness exists with biological developments. In the beginning, the High Yielding Varieties of the Green Revolution seemed to be a panacea. It was assumed that since farmers everywhere knew how to sow, they would know how to do this also with the new seeds. What the planners forgot to take in account was that these new varieties demanded more optimal conditions of growth than the local varieties. They were often planned for large farms using suitable machines which were not in the hands of the peasants. Thus lack of technological and economic capacity led in many places to a failure of the Green Revolution (Dahlberg, 1979).

Another step which is often overlooked in the transformation process of biological knowledge for the use in community developments is the identification and proper use of suitable communication: means by which to reach those who have to work out, promote and decide upon the introduction of an innovation and then to monitor and evaluate its implementation. Here mainly the informal educational system with extension services in the lead, using mass and other media, come in. So far, most of extension is based in giving prescriptions. A major educational effort is needed to educate community leaders how to cope with the new knowledge and how to decide for themselves what is best for them.

Whenever a new piece of knowledge, a new technique, a new approach is transformed from research reporting to use in community development, some of the problems remain for lack of an optimal solution, and new needs are identified. Therefore the search for relevant and applicable knowledge continues and a new cycle starts. As more data on the community and its problems and more knowledge become available and are carefully evaluated, the chances for better solutions to be found grow.

A Case Study

In order to show how the model works in practice, let us follow a case history. The community—farmers in the semi-arid, saline Arava valley in Southern Israel—is climatically and geographically typical for many agrarian communities living under semi-arid and close to arid conditions. Sociologically the Arava community is probably unique, because of the high educational standard of its members. This mixture of common and uncommon features can be found in every case study. Consequently, in case studies some aspects are generalisable, but others are so unique that no ready-made solution can be transferred from one case to another.

The Arava valley stretches from the Dead Sea to the Gulf of Aqaba. It is a desert with occasional winter rains averaging less than 50 millimetres a year. Temperatures reach more than 40 centigrade in August and fall to 17 degrees in January. Relative humidity is very low, and evaporation rates very high. The soils are typical desert formations with wind-blown coarse sand and heavier losses. They are highly saline and have to be leached before they can be cultivated. Occasional winter flash floods originating in the mountains to the West are too brief to allow surface storage, but they do replenish the brackish aquifers. When the water is pumped into furrows for gravity irrigation, less than half of the water reaches the plants, much is evaporated and therefore the water reaching the plants is extremely saline. Erosion is another problem. When sprinklers are used, the erosion problem is smaller, but evaporation rates are even higher, under the extremely dry conditions, and the saline water damages the foliage.

The Arava valley had two big advantages, when cultivation efforts started. Land was available, and because of the warm winter, a high potential for the winter crops existed if suitable varieties and a better-suited irrigation method could be found.

A community description consists of natural-environmental and socio-economic parameters. While the arid conditions described are rather typical for many desert and close to desert regions, the human population which tried to master the environmental problems of the Arava valley for the first time, was rather unusual. The pioneering settlers were dedicated to their challenge, had themselves a fairly good education, were inclined towards experimenting, and had access to the research community. Therefore it was not difficult for them to define what relevant biological knowledge was needed. Either a halophytic crop had to be found and had to prove itself economically feasible, or a technology had to be developed, which would prevent the evaporation of most of the irrigation water.

Still in the 1950s, an experimental desert garden was established near Elat, at the Southern end of the Arava valley, and dozens of cultural plants were tested in order to identify which varieties would be most resistant to salt damage. The most promising crop was Sisal, but its cultivation had no chance to become economical.

Then the second approach—to make better use of the water in irrigating winter crops—was considered. Now the problem became one of technology. For decades the principle of drip irrigation was known. Basically, it is very simple. Water is brought in pipes close to the plants and is then released through small holes in the pipe directly into the area above the roots.

The first request for a patent for an irrigation unit resembling the dripper was filed in California already in 1874. It proved to be premature. About the same time, German farmers laid clay pipes with open joints into the soil in an effort to combine irrigation and drainage, as the water table in the soil rose or fell during the year. It did not work. Another crude method was used in Australia, in the 1930s. Farmers cut holes with chisels into iron pipes to let water dropwise out. This method too did not give the expected results. In 1934, scientists at Michigan State University suggested trying a canvas hose, as used by fire brigades, but without the rubber lining, so that small droplets of water could perspire through the hose. At the same time, in England and Holland, growers tried to use in greenhouses rubber pipes into which small emitters were fitted.

The idea of the emitter was a big step forward, but emitters were too often clogged and the rubber soon became brittle, especially when exposed to the sun. Mice too seemed to like to add unplanned holes into the rubber pipes. To overcome the sun's effect on rubber, Israeli farmers tried to bury the rubber pipes underground, but the roots of the plants infiltrated the unsuitable emitters.

The next step in irrigation technology came after World War II, when the aluminium industry, which had boomed during the war and was looking for new outlets, started to produce cheap pipes which did not rust and could be transported easily. Sprinklers, mounted on aluminium pipes, revolutionised irrigated agriculture, but were not suitable for arid areas with saline water. The emitted drops had a very high evaporation rate, before they reached the soil or the plant. The real breakthrough came only with the uprise of a new technology: plastics.

Some 45 years ago, Simcha Blass, an Israeli water engineer, observed that a large tree near a leaking faucet exhibited a more vigorous growth than the other trees in the area. This led him to the concept of supplying water to plants drop by drop. He developed and patented a low pressure system for delivering small amounts of water to the roots of the plants at frequent intervals. With the development of plastics technology this became possible. Lasers helped to bore very fine and exact holes into the plastic pipes.

The breakthrough would not have come about without the active co-operation of the Arava farmers who had a vital interest in the new development and other farming communities who had also water problems, but also experience with irrigation. One of the communal Kibbutz settlements in the area had a sprinkler factory and was interested and knowledgeable both in semi-arid agriculture and in irrigation technology. This close co-operation between technologists, manufacturers and clients accelerated the further development of the system, mainly through the introduction of a coiled emitter that prevented clogging and reduced discharge pressure. Using drip irrigation, a field trial in the Arava produced a harvest of 58.3 tons of tomatoes per hectare, compared to 35.8 with sprinkler irrigation, drip irrigation proved to be also labour-saving and could be used to distribute water-soluble fertilisers directly to the root system.

According to the "Biotechnical Knowledge for Rural Development" model described above, the next test was economic feasibility. In spite of the high costs of the drip irrigation equipment, it proved to be economical, because the high prices paid for winter vegetables on the European market covered the investment, for which relatively cheap credit was available. Labour costs were saved, because pipes did not have to be moved and could also be used to apply fertilisers.

In the case of the Arava farmers, the problem of information flow through communication channels did not exist. As we have seen, these farmers actually took part in the creation of new knowledge and gave high priority to the implementation of the innovations. Success came and the desert community flourished.

As in any rural development process, new problems came up, new needs were identified, and the cycle started again. With the expansion of the irrigated area, in the Arava valley, water became more and more rare. Varieties with less transpiration and a higher growth/water consumption rate were needed, and the use of water had to be made more efficient. The University of the Negev and other research institutes specialising in desert biology intensified their research on halophytic cultural plants. Also on the technological front, the new needs triggered further developments. Computers which receive data on soil and climatic factors (some directly from measuring instruments in the field) now open and close valves at the right time and help to save water. This quite expensive device proved to be economical under Arava conditions, where the alternatives were water transport over hundreds of kilometres, or a search for deeper water sources, the availability of which was most uncertain.

The biotechnical knowledge which was crucial in the development of the Arava community might not be so useful in the case of other arid zone communities, even if the ecological data match, and the same biological principles rule plant life and can be applied to plant management. Drip irrigation is a quite highly developed technology. Careful planning and handling of the system is necessary. The installation of the plastic pipe system costs per hectare more than other irrigation systems. It is labour saving, but in most developing countries, and especially in peasant communities, labour is amply available, but money for investments and technical maintenance services is not. It remains doubtful if drip irrigation gives the optimal solution under these conditions. A careful consideration of all stages in the Biotechnical Knowledge for Rural Development Cycle is necessary to come up with the optimal plan of action in each case. Only very seldom, solutions can be transferred from one place to another, without adaptation.

References

Benor, D. and Baxter, M. (1964) Training and Visit Extension, Washington, D.C.: The World Bank.

Blum, A. (1982) From biological knowledge to community development. In Atchia, M. (Ed.) Research in Community-based Biological Education. Réduit and Southampton: International Union for Biological Science.

Dahlberg, K. A. (1979) Beyond the Green Revolution—The Ecology and Politics of Global Agricultural Development. New York: Plenum Press.

Frisch, K. von (1946) Die Sprache der Bienen und ihre Nutzanwendung in der Landwirtschaft. *Experientia,* II: 397–404.

2

Information Technology Education in Hong Kong

NANCY LAW

Hong Kong Science Teachers' Association

History

1.1. The introduction of computers into schools in Hong Kong started around 1980 in a few pioneering schools. These came as a result of enthusiastic efforts of some schoolteachers, and they started as interest groups in extracurricular activities. *The computers used were either bought from the school's own funds or were the personal properties of the teachers involved.*

1.2. Though the Education sector (schools, teachers and University lecturers) has put pressure on to the Education Department to finance the introduction of Computers into schools, the most effective push that gave the much needed momentum to go ahead came from the Industrial and Commercial sector, as they are very much worried that Hong Kong would not be able to compete with other countries if we do not go ahead in training our students to use computers in schools. As a result, the Government gave the green light to finance 30 pilot schools (selected from schools in the public sector) to start Computer Studies as a formal school subject in F.4 in 1982.

1.3. Thus, unlike many other countries where Computers started to be used in schools as an educational aid/tool for the teaching of subjects already in the school curricula, we began with Computers used essentially for the teaching of a new subject: Computer Studies.

1.4. The first draft of the Computer Studies teaching syllabus was completed in 1981 and the subject started to be taught in schools in 1982. The objectives of the subject were to enable pupils to understand the functions, uses and limitations of computers, to provide an opportunity for the study of the modern methods of information processing, and to encourage an understanding of the implications of computers in the modern world and to prepare pupils for further studies in Computer Science. The syllabus include:

(i)   Evolution of Information Processing

172

(ii)   Introductory Computer Concepts

(iii) Flowcharting

(iv)   Programming in BASIC

(v)   Input/Output, Coding and the Storage of Information

(vi)   Data Processing

(vii) Computers in the Modern World

(viii) Computer Operation

1.5. The financial commitment that the Government has put itself into in the provision of computers in schools in the last 4 years was huge compared with the overall budget cuts in every Government Department, the Education Department inclusive. Phase One of the Pilot Scheme in 1982 was granted a total budget of approximately US$400,000. This scheme provides 8 sets of microcomputers each complete with a disk drive and colour VDU, of which one was fitted with a dot matrix printer; one set of microcomputer fitted with a dual disk drive and a dot matrix printer for each school. Thirty schools from the public sector. Phase Two of the Pilot Scheme was held last year and another group of 75 schools were selected for this.

As a result of the successful implementation of the Scheme and taking into account the interest of schools and pressure from other sectors, the Education Department decided that it should be expanded so that all public sector schools would be able, if they so wish, to include this subject in the curriculum. The full-scale implementation is expected to materialise not later than 1986 if funds are available. Funds have already been granted for another additional 105 schools to implement this subject in September 1985. (Hong Kong has approximately 400 secondary schools in the public sector.)

Review

2.1. The generosity of the Government in the provision of computer hardware allowed students to have easy access to the greatest technological invention yet since the appearance of the steam engine. Yet such endowment has not been as fully utilised as can be hoped for. As yet, the emphasis has been on the training of students to write BASIC programs, the importance of which is doubtful. With the advent of the fifth generation computers in sight, we are sure that more people will not be writing their own programs, nor will such knowledge be necessary. Rather, what is important for students is to know in what ways computers are being used and how they may be helpful to every aspect of our lives. What is equally important is to know the social implications of the use (and misuse) of computers (and information for that matter). To educate our students effectively in these aspects we need more than just hardware—we need suitably designed software and informed, well-prepared teachers.

2.2. Another huge potential that we have not yet tapped is the use of computers as an educational tool. Again here, we face the same two problems—the need for good software and prepared teachers. Many teachers find the computer a formidable machine whereas few students would share such sentiments. Thus we must educate and train our teachers in the use of this newly acquired technology. Furthermore, we need more research into the potentials of this tool in education—e.g. how may computers be used to promote the cognitive development of students?

Microelectronics in Education

3.1. Taking into account of the above problems, a new committee entitled the "Microelectronics in Education Committee" has been established under the ambit of the Curriculum Development Committee of the Education Department. The main function of this committee is to consider the development of microelectronics throughout the school curriculum.

3.2. The first meeting was held in December 1984. The committee generally felt that the programming part of the Computer Studies syllabus should not be over-emphasised and the syllabus is currently under revision.

3.3. Members also considered that computer awareness was essential for youngsters. However, it was not the right time to start the subject in primary schools at present because of the lack of trained teachers. They hoped that more computer courses would be organised for primary teachers.

3.4. Members were concerned about the high cost of computer software and felt that large numbers of CAI/CAL packages might not be afforded.

3.5. Furthermore, CAI/CAL packages developed in other countries may not fit our local situation, and there is a need for locally developed educational software.

Computer Education Centre

4.1. It was announced in December 1984 that a Computer Education Centre will be established under the auspices of the Education Department.

4.2. The capital expenditure of the Centre will be $4m and the recurrent commitment estimated at some $480,000 for the first year.

4.3. Conceptually the Centre will serve both as a training and resource centre. As a training centre, it serves as a venue where courses of various natures for both Computer Studies and non-Computer Studies teachers will be organised. As a resource centre, it provides facilities for the exchange and distribution of CAI/CAL packages. There will also be a library stocked with reference and latest audio-visual materials for use by teachers. It will consist of three areas, namely the Resource/Demonstration Area, the Teaching/Training Area and the Reference/Support Area.

3

Microcomputers in Arab Education: Problems and Prospects, the Case for Jordan

GHASSAN F. ABDULLAH

Jordan

PARTLY THROUGH pilot projects at Jordanian schools, this paper outlines the main problems and suggests guidelines for the introduction of computers in Arab education.

Introduction

The student population in the Arab countries is estimated at 35 million in 1984–85 and is expected to reach about 65 million students by the year 2000, if present trends continue. It is still far short of a place for every school-age child. Illiteracy rates persist at over 60% in some of the more populous Arab states. In Jordan, with a population of 3.1 million, illiteracy stands at about 40%.

Where do computers and high educational technology fit into this picture?

The commercial potential of the Arab school market has not escaped many western and far eastern computer manufacturers. Competition is driving prices down and improving specifications, including "Arabic". Many private schools have started computer classes to cater for the wishes of parents, especially among higher and middle classes, to have their children computer literate. The promise of computer education and computer aided learning is also being considered by some education ministries in the area. The advantages of this new interactive educational tool cannot be ignored. The few dissenting voices warning of the disadvantages, centring mostly on the "inhuman" aspects of computer education, have been drowned in the chorus of acclaim for CAL. The questions argued will no doubt be with us for a long time to come. They will not be dwelled upon in this brief. Only the particular problems concerning the introduction of microcomputers in Arab schools, through the case of Jordan, will be outlined.

Present Situation

The introduction of computer education and CAL in Arab education is facing many problems that might be grouped as follows:

1. Traditional educational methods in the Arab World tend to give a greater role to the teacher, do not encourage critical questioning and put too much stress on performance at examinations. Introducing new subjects and developing older ones is very slow. All this might be at variance with the newer methods of CAL such as those based on simulation. A questioning and exploring attitude is natural in the interactive computer environment, even with the tutorial and drill and practice techniques. Scolding for mistakes is administered much more lightly by the computer. A lot of rethinking into present educational methods has to take place before CAL gains wide acceptance.

2. Arabisation offers a challenging problem to computer manufacturers, as is the case with many other languages. Writing from right to left, the different forms taken by characters according to their position in the word, the vowel signs added on top or below letters, the use of "Arabic" or "Indian" numerals and the additional problems of simultaneous bi-lingual needs, are all points to consider. Many solutions, hardware and software based, already exist. The Arab Standards and Metrology Organisation (ASMO), one of the Arab League's organisations, has drawn up recommendations for standard codes (ASMO 499/1982). This Arabic Standard specifies the properties of a coded character set using 7-bit binary codes for information exchange among different types of data processing equipments using the Arabic characters, to quote ASMO. Many computer manufacturers and software developers have felt the need to bypass the ASMO standards or go it their own way about implementing Arabic. A more comprehensive and "binding" standard, such as ASCII, is clearly called for.

3. The fast growth in student numbers in most Arab countries, reaching about 4% annually, is just ahead of the overall population growth. High illiteracy rates will, therefore, persist in the foreseeable future, in spite of the literacy campaigns. In Jordan, illiteracy stands at 40%, although on any school day, about 30% of the total population sits at study benches across the country. Education budgets are further stretched by the need for more qualified teachers, better books and other educational material.

The introduction of microcomputers, on a wider scale, might have to wait, perhaps, until unit prices drop to as low as $50 per station for the benefits of computer literacy and CAL to reach more than the privileged few in the Arab World. Microelectronics, strange as it may seem at present, could even hold a promise in illiteracy fighting. The computer would go to the people instead of the people going to a literacy centre. This arrangement would be particularly suitable to women. They constitute more than 70% of the total number of illiterates and are usually burdened by all the domestic and other work.

Advances in hardware and software such as ease of use, alternatives to the keyboard, enhancements in sound and graphics and other developments could help planners and developers of literacy campaigns develop suitable programs using computers. Clearly, all this is pure speculation at this stage. More studies are needed to look more closely at the feasibility and problems involved.

4.   The introduction of computers, so far, in most Arab enterprises, both public and private, cannot be described as a resounding success. A recent official study, undertaken by a special government committee in Jordan, points to gross inefficiencies and under-utilisation in the use of computers in the public sector. Out of a total sum spent on computer systems by the government of 7 million Jordanian Dinars ($17m), over 80% was spent on hardware and only 11% on software. Training got less than 1% while the shortage of qualified and trained personnel is decried. The situation in the private sector is only marginally better, but no detailed studies are available.

If the introduction of computers in education is to be spared the same fate as in the Arab public and private sectors, what should be done? To begin with, it should be recognised that people take priority over hardware and software. More computer knowledge and appreciation should be made available at all levels, from public officials at ministries of education to the all-important teacher level, within overall national plans for computer education.

5.   Development of Arabic educational software is hardly started, most of it so far being a translation of existing western material, which in turn still leaves room for improvement on the whole. Some indigenous work is already being developed locally, often with the help of foreign affiliates. Of particular note is the Kuwaiti al-Alamiah effort, based on arabised MSX, which is producing interesting and well-thought packages.

A market for school syllabus courseware is already opening up, prompted by calls from education ministries in some Gulf states, north African Arab states and others. In Jordan, the ministry of education is calling for tenders to produce courseware for the mathematics syllabus for the first secondary grade.

Current Education

Computer education is taking varied forms in Jordan and the other Arab countries at present, and at different levels. Pan Arab "summer camps" for computer literacy and other activities, are taking place this year in four different locations, catering for the children who can afford them. Dozens of private schools, child centres and commercial schools, are running quick computer courses, taking advantage of the interest generated by this prestigious pursuit.

On a more formal level, professional formation of computer science and engineering graduates takes place at the Jordan and Yarmouk universities. Two and three year diploma courses are also offered at intermediate higher education institutes, the Community Colleges. However, finalists find it harder to get employment, due mostly to poor English and initial deficient recruitment standards, according to employers.

Of particular interest in computer education, is the pilot project conducted by the Jordanian ministry of education, based on a plan of action drawn up and approved in 1984. It was started, in the fall of that year, at two Amman public schools, one for boys and one for girls, with 11 micros each in a computer lab. The experiment is still being monitored and will be extended to 6 new schools in the provinces, with about 90 micros in total. It is too soon to draw full conclusions, but some problems encountered so far, also at private schools conducting similar projects, could be summed up as follows:

—To find and train the right teachers.

—The necessity of English. In the future properly designed English instruction using CAL methods could be used to rectify poor English.

—The resentment of some teachers at seeing young students becoming quickly more fluent at using the machines, thus robbing the teachers of some of their prestige as know-alls.

—The lack of sufficient facilities to cater for large numbers of classes and students at the computer lab.

—The no less important organisational, technical and administrative problems.

Lessons, of course, are being learnt, and the availability of more funds, and Arabic courseware, would advance matters further.

Prospects

The Arab education systems are far from unified, but they share some common concerns such as the need to expand further, to develop and modernise. The introduction of computers in education is posing new challenges, and opportunities, for educationalists and computer people alike. A few guidelines might be suggested to tackle the common problems and ensure a smoother transition into the information age, the education system being one of the most important agents for this transition.

1. At the national level

The introduction of computer aided education has wider scope and implications than could be handled by education officials alone. Other parties should be involved such as the universities, the council for educational technology, the ministries of education and higher education, computer education consultants, etc., in addition to parties involved with overall planning and development in the country. A national plan would be drawn up to formulate general policy, lay down detailed steps and institute follow-up procedures. Such a plan would take stock of actual needs, try to avoid the pitfalls of computerisation in other disciplines and lay special emphasis on the training of the people who will have to carry it out. It would also draw on the experiences of countries already ahead in this field.

2. At the pan Arab level

The common problems in Arab educational systems and the technicalities of Arabisation should prompt a concerted Arab effort, through various specialised organisations of the Arab League and national bodies concerned, to develop a more unified approach to those problems. In particular, the ASMO Arabic codes could be made more comprehensive and binding. Applying high technology in the fight against illiteracy could also be looked into on the Arab level.

3. At the international level

Arab and other countries could benefit further from an exchange of information, through conferences and publications, on computer education and related subjects. An international body to turn to for help and advice on the formulation and execution of national plans for the introduction of computers in education, and in the detailed choice and evaluation of products, would serve present and future educational needs of many a nation, particularly developing nations.

4

New Information Technology—Quo Vadis?

MAURICE EDMUNDSON

*Former Staff Inspector at the Department of Education and Science with special
responsibility for the Microelectronics Education Programme in schools. The paper was
originally written for, and published by, the Independent Schools Microelectronics Centre
in Oxford, England.*

1. For many schools the acquisition of a language laboratory became a prime aim. The Heads of those schools which were fortunate to have them first were recruited into the Inspectorate so that this development could be monitored by those with experience of them. Ten years later I witnessed many of these laboratories being torn out and I recall one institution (a university as it happened) where the tape decks were offered to other departments in the hope that they might make better use of them. There are, of course, installations at all education and training levels still in use. It was the mass-movement especially in the schools, which failed.

2. In the 70s educational technology was promoted in other ways: resources for learning; distant study; individualised learning; structured worksheets in classroom and laboratory; and so on. The 80s is the decade of the micro, of software and interfaces, of data storage and retrieval and electronic communication. How will they all be viewed in the 90s? It would be wrong to suggest that these strong surges of innovation inevitably recede back to the same starting point. Some gains are made and, in our attempts to improve the system, we edge forward by small but hopefully real amounts. The question we should ask is "what lessons have been learned from the 60s and 70s movements in educational technology which may prevent us from repeating mistakes with the new technology in the classrooms of today?"

3. So far as the hardware is concerned, microelectronic devices are many times more reliable than their electro-mechanical predecessors. Microcomputers and other associated technologies are not by any means free from hardware problems, but these have diminished a great deal since the worst days of the language laboratory. If lessons are to be learned from past educational technology we have to be aware of both the pedagogical issues in today's curricula as well as the organisational and methodological constraints for both teachers and pupils, and apply our knowledge to the use of the new

180

information technologies in the classroom. Based on my own experience of educational technology developments over the past 20 years I single out the following as important for consideration in the microcomputer age:

1. Teacher–teacher and especially teacher–pupil relationships.

2. Software issues.

3. Organisational matters in the school and the classroom.

4. A very large slice of the vocation of teaching is concerned with personal relationships—with colleagues, with pupils, with parents and so on. Every reasonable teacher has experienced that immensely rewarding feeling of satisfaction and achievement resulting from a superb interactive lesson with a class of children. If only there were more lessons like that; thank goodness there are some! I see these relationships as essential in the year 2000 as they are today, and therefore I see the importance of the micro in schools mainly in that context—how they can improve the quality and the extent of teacher–pupil and teacher–teacher interactions and relationships. Microcomputers are valuable tools; they will not replace teachers in the foreseeable future. We strive to make programs more user-friendly, by which is implied more human. But a machine is a machine is a machine; a computer's response is pre-ordained by its program however sophisticated it may be. In the right context, one raised eyebrow has still more thought-provoking power than a screenful of computerised questions.

5. Although educational computer programs have been with us for some time there has been an explosion in output over the past 2 years, which shows every sign of continuing to grow. Sadly, a fair proportion of this software is at the level of the bad programmed learning of the 60s or worse. Extravagant claims are made (especially in the home-education market) about motivation, learning is easy or fun, etc. Misleading statements imply that all the shortcomings of teaching and learning in past generations of pupils can now be overcome by microcomputer programs. The seeds of distrust and disillusion which finally undermined the new technologies of the 60s and 70s are already being sown in the micro age of the 80s. This is a pity. The micro is such a valuable, even unique aid, that the good uses must not be dragged into disrepute by the bad.

6. This raises the question "what then, are the good uses of the microcomputer?" I have already indicated my belief that "good use" will always enhance teacher–pupil relationships. "Good" software will assist the teacher to do a better job and thereby increase vocational satisfaction and self-esteem. Teachers are very individualistic and as with all other teaching aids, even textbooks, one teacher's ideal is anathema to another. For this reason lists of good and bad programs have limited validity. For example, within the varied, daily regime of individual and group activity to be found in many successful junior schools, a simple program for table practice or number bonds may fit perfectly well. There may be other junior school environments where such a

program would merely add to the arid diet already on offer. Software has to be judged in context.

7. Thus it would be wrong to make a blanket condemnation of all drill and practice type programs. In the right context they may be comparable with a textbook exercise, perhaps more exciting; they might release teacher time and energy for more challenging needs, and so on. Condemnation may be wrong but caution is essential. For example, good teaching builds on misunderstanding and mistakes, as well as on correct responses. Most programs are a long way away from being able to react to a student as a human being would. In a survey looking at the use of microcomputers in the teaching of mathematics, conducted in 1983, the author writes:

A general advantage of computer-managed drill and practice is that anxiety is largely absent. At its best, computer-managed drill and practice enables the child to compete with himself, with tasks matching his level of achievement, with rapid feedback, and with a confidential record of his scores available to the teacher at the end of the day. But if practice is to be of value, needs have to be diagnosed, and there is no point in indiscriminate practice which is unrelated to the child's level of development. In these matters the teacher has to be the judge; as yet no software has been seen assisting in this role to any significant extent.

8. In 1983 I organised a conference of educationists of considerable experience and reputation in their own subject field; we attempted to analyse in what way the microcomputer could enhance learning in the school curriculum in geography, science and mathematics. The outcomes were valuable, providing much useful information about the type of software we should be devoting our limited resources to produce. This planned approach to software production seems more sensible than the arbitrary way in which software has been written to date. Unfortunately I do not think it is likely to happen. Even within a nationally sponsored programme, such as the Microelectronics Education Programme, it will be difficult, but in general I expect the published output of microcomputer programs to be a random affair. There is already some evidence to show where this will lead. There will be considerable duplication, many programs attempting to do the same thing with perhaps some slight variation. Much of the software will reflect the methods and attitudes of authors, whose traditional approach ensures the status quo; the truly innovative programs will be in the minority. There will be little incentive to improve standards (that is, computing and educational standards) beyond a certain market level. This will lead in my view to a drop in the use of computers in the classroom. By the end of the decade the pendulum will have swung back towards its starting point, just as it did in the 60s and 70s. Of course, teacher training, both in-service and initial (indeed, initial most of all) has an important part to play in preventing the swing from going too far.

9. Teachers usually agree that their main job is to prepare their pupils for life, not just for examinations; to train them how to learn is as important an objective

as learning the details of a relatively narrow syllabus. Social pressures work the other way. Examinations are important and priorities become reversed, however well intentioned the teacher may be. Long-term aims are frequently sacrificed for short-term goals. Memorising the algorithm (in whatever context or subject) which will improve the chances of examination success is often "safer" than slogging away for the conceptual understanding which would last a lifetime. I believe that I have glimpsed within the microcomputer a tool which will narrow the gap between these conflicting pressures on teacher and pupil.

10. It is for these reasons that I support most strongly those programs which assist with conceptual understanding. If I were a history or a geography teacher, perhaps I would achieve this through open-ended simulation type programs. In science whilst experimental simulations may also be helpful, many instances come to mind where the micro can support a "conceptual" model more effectively than any other aid. The mathematics group at the conference, to which I referred above, produced a paper stressing the value of direct interaction of pupil with computer in which the learner explores for himself algorithms, algebraic relationships, geometrical properties, functions, etc. Programs would vary from a few lines (or even none at all!) to the more conventional sophisticated package. I think these ideas should be extended to many areas of the curriculum. The power to use the computer (in the sense of bidding it to do one's will) has been neglected. Past educational technology still permeates the system and it tends to direct our energies into forms of CAL which are too close for comfort to the electronic equivalent of programmed learning.

11. I have no qualms about young children learning to "program" the computer (if that is the right word), not with the intention that they should write formal programs themselves, but because it is a fundamental and essential skill if the pupil is to master the computer and use it as a tool on which to sharpen and develop his/her own intellect. Some keyboard-finger mastery would also be desirable. Speech input as sophisticated as the keyboard is a long way ahead. Keyboard skills for computer dialogue, word processing, telecommunications via the micro terminal, will be valuable for some time to come. A child of 7 could learn these skills in a few weeks of daily practice and they last forever.

12. One further conclusion is that computer languages may need a good deal of refining for maximum effectiveness in the roles which I have been advocating. Amongst the other languages which may have special value in schools, the limited knowledge I have of LOGO suggests to me that this is a powerful language likely to be of value as schools come to terms with alternative uses of the micro, particularly in the first half of compulsory education. I refer of course to the "full" version of the language, of which turtle graphics is a part. The question of language for use in schools is a topic which we have hardly yet begun to explore. We are just beginning to glimpse trends in software design which will make knowledge of computer programming in any specialised

language unnecessary for most people. A level of human-computer dialogue will be possible before the end of the decade which will increase many times the value of the computer as a flexible learning aid.

13. If my views about computer use in education are even partly realised, the impact which they will have on classroom and school organisation is considerable. At the time of writing, almost every primary school has at least one micro. The "typical" secondary school has 8 or 10 micros. An increasing number of schools have a computer room fitted with 15 or more. These are linked as a network, or will be so linked as money becomes available. The wheel has turned full circle since the first language laboratory was installed. Schools will continue to purchase more microcomputers in the next few years, even though subsidised schemes are withdrawn. Costs will fall and schools may be tempted to buy cheaper machines. This is no bad thing for the type of individual interactive use I have been describing, providing a reasonable degree of compatibility with the school's software stock is maintained. A school may therefore have both networked and individual micros. The network may extend around one room, or across several. A mathematics laboratory for example, may be on the network with three or four of its machines but will also have many more simple systems for direct individual use giving the same degree of flexibility as the electronic calculator does today.

14. As for the ever more advanced electronic systems just around the corner, where micros are linked to projectors, tape recorders, video discs, and so on, I am content to let them be developed by those who enjoy doing so, without feeling any particular enthusiasm about their value to schools. They may help the physically handicapped; they may be of use in vocational training in industry and commerce (where many educational technology systems like programmed learning still successfully survive), but I cannot imagine them in school classrooms in the near future.

15. Where then are we going with micros in schools in the next 10 years? In the UK we have been fortunate to have a strong government commitment to NITs in schools, which has moved us forward a good deal faster and further than most other European countries. The MEP is well designed and well run. There are plans for it to continue until 1986 at least. This means there will be continuing coordinated effort with software production and with teacher training, whilst the information and other services provided by the Regional Information Centres should continue to persuade LEAs to keep in touch with each other and encourage more of them to back up the regional service with additional facilities of their own.

16. By the late 80s the majority of secondary schools will have 8 or more micros, some or all of which will be used in areas of the curriculum other than computer studies. In many schools these micros may be linked in some form of network. The use of computers in the curriculum as an aid to teaching will be relatively small, largely confined to geography, some history, mathematics, a little science and home economics. There is some evidence that language teaching will be making use of computer facilities too. Most primary schools

will have 2 or 3 micros; the pattern of use will vary widely, depending on the enthusiasm and skill of a few members of staff. Languages other than BASIC, for example LOGO, could be making an impact in junior and lower secondary classes, but a shortage of trained teachers will limit the most imaginative developments in this field.

17. In very few cases will the use of the computer, however good the software, be other than a cosmetic addition to the teaching. It will take much longer for the unique qualities of the micro to be properly exploited by both teacher and pupils especially by teachers of subjects outside mathematics and science. Electronics (within science courses or craft/design/technology courses) will continue its slow growth, although changes in examination syllabuses could speed the process up as could (in the UK) the influence of the new training initiative on the vocational aspects of the curriculum. In laboratories, self contained programmable micro-processor devices like the VELA and the GiPSI will become more commonplace and there will then be less need to devote a micro for data capture and display. In laboratories and workshops, the micro as a control device will also be more frequently seen and word processing and text editing will be new, if sparsely used, tools in office and business studies courses or in English and other language departments. I do not see Teletext/Viewdata type systems being much used in schools in the UK by the mid or late 80s, although simulated in-house versions could be a useful and popular teaching aid, once the techniques are known.

18. Predicting the future is a risky business; much of what happens with the development of new technologies in schools will depend on the level of investment afforded. Pump-priming is a useful technique but longer term investment is sometimes required if the innovation is to be truly grafted on to the educational system. I believe the new technology will require attention on at least two fronts, if it is to grow and endure. The first is an on-going commitment to teacher training and re-training. The second concerns the realities of software production. If I am correct in believing that teachers want, and will use most of all, educational "utilities", teaching "tools" and materials which assist the development of concepts rather than merely support the factual side of learning, present evidence suggests that publishers are unlikely to find this side of their work very profitable. How then will the output be maintained? Finally, a problem well known to science teachers is that of enthusing girls for the new technology and of meeting their particular needs and interests at an important formative stage of their development. I hope microcomputers will open up in the next few years, many new routes for equalising interest and opportunity between boys and girls. Not all experience so far is encouraging but the challenge is real enough and one I hope we shall win.

References

Fletcher, T. J. (1982) *Microcomputers and Mathematics in Schools*. Department of Education and Science.

5

The Interface between Computers and the Naive User

ANNE LEEMING

Centre for Business Systems Analysis,
The City University Business School

COMPUTERS of all sizes have been with us now for over 30 years and it is now almost unthinkable that we should not use them. They present a challenge to us to harness their power carefully for use in many different problems and in applications which will bring benefit to the user rather than creating bigger problems of bewilderment and frustration. To provide an encouraging environment for the successful use of a computer it is essential to plan for its arrival and control its use.

Modern Microcomputers

New users are most likely to meet a microcomputer, a machine that can cost from a few hundred pounds to a few thousand pounds. It is capable of very much more than its physically bigger predecessors and at a fraction of their costs, both capital and running. Since the new system is small, attractively designed and compact it looks as if it might be approachable and easy to use. It comes with a reputation for greater reliability than its bigger siblings and in many cases it is more powerful and more sophisticated. Many versions of these machines are portable, needing little more effort to carry around than a domestic sewing machine. Some are very small indeed, even pocket sized, and find use on factory floors or in salesmen's pockets, as well as for amusement and educational purposes in the home.

Software

The microcomputer hardware on its own is of little use to the new user; it needs to be provided with two kinds of software; these are systems software and applications software. The former kind is machine specific and is usually the provider of the first interface for the user. Choice of this category of software, or the OPERATING SYSTEM, is extremely important. It is where the user, after some experience, will need to operate to create security copies of software and do other essential housekeeping activities. Fortunately there has been a spate of new products of this type recently so the user can have an operating system nearer his needs than was possible a few years ago. The second major category of software, APPLICATIONS SOFTWARE, provides the user with a set of instructions for a particular need, such as an exercise in mathematical skill or a geographical case study. The end users' needs are first met at this point, whether in educational work or in business tasks or in any sphere.

The Operating System

The microcomputer, when it comes from the supplier, has some software (a set of programs) with it that has been designed to execute some very standard tasks within the computer. Some examples include the testing for errors in the hardware, transferring data between the various parts of the hardware such as the screen and disk, loading and saving of data and program files from peripheral units, naming of programs, simple editing of saved information, copying of files, deleting old files and many other basic tasks. This set of functions is contained in the operating system. These may vary considerably in what they offer to the user in the way of facilities and in how easy they are to use.

Applications Software

The microcomputer system can be put to any number of tasks; most of which are probably not yet invented. The common tasks, nowadays, include teaching aids, both in the classroom and over long distances using networks, administrative and operational tasks in small businesses, use as desktop aids to executives in larger organisations. Such systems are capable of being linked to other small machines or to mini and mainframe computers. The new user is likely to meet the computer in any of these applications. It is also very likely that the new user is going to want to develop his or her own applications, once a first system is in use and needs development. A common and very wide ranging application in nearly every field is word processing; this can be a particularly good application for the new user to start with as existing skills are not displaced. They are enhanced by the removal of tedious and boring aspects. As well as form and letter writing word processing can be used for developing teaching materials in combination with graphics packages which have reached the market place in quite large numbers. The motive for use is probably not terribly important as the skills that need to be learnt on first meeting the computer do not vary much from application to application. There is a tendency for the most recent applications software to reach the market to be developed as turnkey systems; here the user is given simple instructions to begin using the system and then every action needed is prompted from the screen, no consultation of documents is necessary.

The First Approach

Many people admit to some interest in the computer; successful use, even at playing games, brings a feeling of pleasure, of achievement, of power, of desire to learn more. On the other hand, where a first meeting is unsuccessful the new user can rapidly become disillusioned and frustrated, and turn away from this new tool with hope lost and feeling of great disappointment. Many of these feelings are engendered by the myths and glamour that precede the arrival of a computer; and which conveniently omit to mention that it does require some patience and understanding, not to mention technical backup, before successful operation follows. It behoves us, therefore, to look carefully at the process of introducing new users to computing technology so that the users get the benefit they want. If that does not turn out to be the case then it is crucial that the user understands the process so that, at the very least, understanding of the situation is reached.

The Interface Between the New User and the Computer

This can, for our present purposes, best be described as the introductory stages in the relationship between the new user and the computer system. It is also commonly taken as the environment in which the machine is operated, including such factors as physical comfort and the nature of the dialogue between the systems or applications software; both aspects are important to the new user. The aspect that can be controlled at local user level is the introductory stage so I shall concentrate on the relevant factors.

The new user approaches the machine with a mixture of feelings and skills. Some of the feelings we have discussed, such as interest in something new and high expectations. There is also likely to be anxiety about ability to use the machine, to repair it, and possibly about job security. Anxiety should be understood and removed otherwise learning about the new technology will be impeded or blocked altogether. Anxiety about ability to learn can be lessened by the introductory programme followed by the user. This should proceed at a pace appropriate to the new user, finding out and building on existing skills such as the ability to type and by using simple tasks as an introduction. The organisational position of the new user needs to be clear especially as regards the care of the new machine; the tasks that have to be done include the maintenance and repair of all parts of the machine, the provision of supplies such as printer ribbons, paper, and the correct diskettes, the care of the systems software as well as the applications programs.

Having briefly introduced the main aspects of the interface it is necessary to consider some in more detail. The important ones to consider are the choice of operating system, applications software and the learning programme to be followed by the new user.

Choice of Operating System

It is possible that, in spite of greater choice in the market place, the user may not be given any choice of operating system, at least with the first application. Purchasing decisions, made at a higher level, may lead to the acquisition of a standard operating system for a large group of users. This will have the advantage of standardisation of application development but may, at the same time, deprive some users of the most suitable environment for them. The operating system, in its role of housekeeper/administrator of the hardware system, needs to communicate with the user at various levels. If the application is a turnkey system, then it will not be the application user who will get involved but the system developer and maintainer. They will need the operating system to provide the best environment for development; that is one where program design and debugging is simplified, where the operating system messages are meaningful, where access is secure and yet can be organised for the user's benefit, and where file handling is provided. Using the computer as a tool for education or for developing in-house systems will demand a good and suitable operating system. It is beyond the scope of this paper to discuss the wide variety of operating systems available in any detail; it is sufficient here to point out the areas that need attention and some criteria for selection.

Selection Criteria for Operating Systems

The next section is put in the form of questions that should be asked of the suppliers, by those who provide the new system, after consultation with the new users.

1. Does the operating system provide the following facilities? Copying of files between devices, editing of files, re-naming of files, a full directory of file usage, security precautions, protection features.

2. Is the documentation easy to read, clear, complete, and updated adequately? This is so obvious that it is frequently overlooked; it is particularly important where the computer is going into new areas where there is not a lot of expertise. These attributes are hard to assess quickly as well as being hard to obtain; it is therefore important to ensure that there is available good back-up support in the form of expert humans who are accessible.

3. Examine the messages that show on the VDU screen, are they understandable? With little effort? Does the manual help?

4. Does the operating system support a sufficient number of high level languages for the intended applications? Again, access to other users' experience will provide much needed experience here.

5. Does the operating system allow the user to do more than one task concurrently? This will be important with the more powerful microcomputers such as those based on the Intel 8088 and Motorola 68000 processors. These

process data internally much faster than the Z80 or 6502 based machines but are slowed down by having to use similar peripheral devices.

6. Are the error messages clear and understandable? Does the manual explain what could give rise to the error found? Many error messages are abbreviated on screen in the interests of economy of memory; this can be satisfactory if the manual is good. If a HELP facility is included then examine the text put up on the screen for the characteristics mentioned. The disadvantage of HELP systems is that they take up storage space. If storage space in memory or on disk is critical then HELP text can be printed out and erased from the disks.

7. Is the operating system stored on Read only memory or does it occupy main memory, thus occupying memory otherwise available to the user?

Choice of Applications Software

Many new users will come to the machine with this choice already made. It is not advisable for a new user to make the selection of application software for use where the new procedures will fundamentally affect the work of several people. In many cases, too, choice will be restricted by the availability of the hardware and by other choices previously made. However, there may be instances where new users are drawn into the selection process at an early stage in their experience.

Satisfactory answers to the following questions should be sought:

1. Does the software carry out all the functions required of it?

2. Is there adequate documentation for the new application?

3. Will it suffice for training purposes after an initial demonstration?

4. Does the documentation version match the version number of the software?

5. Is there further human support available with reasonable accessibility?

6. Are all error situations adequately covered in the documentation? i.e. are there full explanations of what to do when an error happens and the likely cause?

7. Is the method of data input to the program clear and controllable?

8. Is it easy to get the required output?

9. Who owns the source code? (The original code in which the system is written.) This is important for the following reasons; correction of errors and future enhancement of the programs. If the supplier still owns the source and goes out of business the purchaser suffers from the fact that the source code in these areas is not available for the user to work on.

10. Can alterations be made to the system without invalidating warranties?

Depending on the detailed nature of the program many further questions could be framed; the above are simply a start to help the user in a shop or exhibition where there is perhaps an unexpected opportunity to acquire software.

Learning Programme

New users have to learn how the hardware system operates; this is the sign-on procedure, how to operate the disks, the monitor, the printer, the tapes and so on. Programs have to be loaded into the processor and data has to be entered. For most configurations this will mean acquiring keyboard skills. Some of the latest systems use touch sensitive screens or light pens or a mouse. This is a device which rolls along the desktop manipulating a symbol on the screen between the various functions that the particular machine is programmed to do. However, these devices are more expensive now than a keyboard. This latter device may appear as the conventional QWERTYUIOP layout with extra keys added in non-standard places or it may simply be a numeric pad. Obviously typing skills are valuable here, however, there do exist a number of aids to teaching keyboard skills which use the computer itself.

The choice of first task is then important. For those with keyboard skills word processing is a suitable place to start. In addition to text entry control commands and editing commands have to be learnt. There are many word processing packages available and the documentation is usually good. The first word processed document is usually produced in a short time, with many of the relevant skills acquired within a few working days. For people without keyboard skills a good approach is to spend a little time playing one of the numerous games written for every machine; this will overcome any latent fears of the technology and rapidly satisfy some basic instincts. For those of a more serious disposition a keyboard tutor is a possible starter or a personal diary package. This is the stage where the documentation is fully tested; if a new user cannot get going along with the documentation on what is claimed to be a turnkey package then the documentation fails the first test.

To summarise the introductory steps:

1.   Learn how to switch on and set up the hardware and all the components; how to load tapes, disks, printer paper; how to switch everything off.

2.   Choose a first task, either a game, a keyboard tutor or a simple application package and follow the instructions accurately.

3.   Become proficient at the first task before choosing a harder one.

4.   Do not be afraid to ask questions; try the documentation.

5.   If all local help fails consult the supplier or previously arranged back-up support.

6.   Enjoy the work; this is the one imperative.

7.   If too many personal mistakes are being made check the physical

surroundings, the lighting, the arrangement of the furniture, the angle and position of the screen and the keyboard, space to hold manuals and paper and pencil, the temperature and dust levels, the height of the chair. Micros are pretty robust, environmental conditions are important for humans.

8. Before too long and certainly before any serious development work is attempted **learn how to make security copies of all software.**
This is VITAL, if it is not done many many hours of work can be lost at a stroke. Everybody does it at least once; even the experts. It is usually due to accidental moves; the implication is, in long sessions at the keyboard, such as in word processing, to remember to save copies of the work as it is done. (I have just saved this text yet AGAIN.)

9. When simple data entry and program loading and execution is a familiar task choose a programming language to learn.
The most likely language to be met on micros is BASIC. This will meet its inventors' objective of getting the new user started quickly; however, once started the subsequent stages are not so satisfactory. If there is access to PASCAL this is a splendid language to use to write programs. Alternatively, look for an Information Storage and Retrieval package; this will have some commands to help file data and access the data needed. Output will probably be in tabular or report format; a graphics package which can take the data from the files and display it graphically is a useful adjunct. Yet another alternative is to go for an application generator. This eliminates the need to write programs but is only suitable for standard data processing applications as its output is usually just textual.

10. Subscribe to a magazine or journal that covers the system being used. Most of these are written for the new user and contain articles written in simple language explaining developments. Alternatively, join or form a user group of people using the same system. This shares experience and solutions to difficulties found as well as providing all the other benefits of being in a group.

By this time the "new user" is an experienced user and is probably training other new users. Experience in the UK shows that the microcomputer user gets thoroughly absorbed in the system, continually wishing to add features and applications to it and keen to share experiences with others as well as acquiring a stream of new experiences. As indicated earlier most of the uses of these systems are not yet invented; many discoveries are there for the making.

Glossary

This glossary contains some of the technical terms used in the text above and some other commonly met terms that may be of use to you in your first encounters. For further help please see the dictionary in the booklist.

APPLICATIONS SOFTWARE: a set of programs that have been designed to perform special functions; *i.e.* an accounting system, a word processing program.

ASSEMBLY LANGUAGE: a coding system that is very close to the machine code of the computer; it is not immediately understandable by the novice, indeed by many computing professionals! It demands special training to use it.

BASIC: an acronym for Beginners' All-purpose Symbolic Instruction Code. A language designed to enable newcomers to write programs with minimal training; it is very successful at this job but does not produce, in general, easy to read and maintain programs.

BIT: the smallest unit in a computer; it can take two values: 0 or 1 and is the basis for all the workings of the computer.

BYTE: the smallest addressable unit in a computer's memory that is capable of holding a character, *i.e.* an A or an H or a 7. For a piece of text such as "it is raining today" the computer needs 19 bytes. Count the letters to check (remember a blank is a character to a computer).

DISKETTES or DISKS or FLOPPY DISKS: the small (5.25 inches in diameter) disks which are coated with a magnetisable surface and used to store data and programs for computers. There are other sizes available.

DOCUMENTATION: the set of manuals which accompany any part of a computer system, the hardware or software, and which explain how it is used, its objectives, functions, what to do when it goes wrong, etc.

EPROM: see under ROM.

HELP SYSTEMS: messages, often written within the application or system software, which will come up on the screen in response to a request from the user and which attempt to explain the function of a command or explain an error that has occurred.

HIGH LEVEL LANGUAGE: a coding system for the programmer which is close to an English way of writing a problem solution. They are easier to use than assembly languages which they have superseded in most business and educational use. Examples include BASIC, COBOL, FORTRAN, PASCAL, ADA, ALGOL, RPG, and many more.

MACHINE LANGUAGE: the coding system that is unique to the particular computer; it is expressed in binary or octal or hexadecimal notation and is almost never used to write programs. Occasionally an understanding of it might be needed to remove a particularly difficult error.

MAINFRAME COMPUTER: the descendant of the original large computer, still in use today in large organisations such as multinationals, banks, insurance companies, weather forecasting, military applications.

MICROCOMPUTERS: the group of computers which have appeared on the market since 1977: they are in the lowest price brackets but are not always less powerful than their bigger siblings. They are most commonly single user systems though multi user systems are now available.

MINICOMPUTERS: the group of computers intermediate in size between micros and mainframes; they are descended from those machines designed for the space effort and are consequently very rugged and reliable in operation.

MONITOR: the component of a microcomputer system which displays the current activity; it is also called a VDU or Visual Display Unit.

MOUSE: a new input device that avoids the use of a keyboard; it rolls about on the desk top and, by means of buttons, allows the user to activate symbols or ikons, on the screen to perform the required function. They need special software.

OBJECT CODE: see MACHINE CODE.

OPERATING SYSTEM: the set of programs that comes with a machine when purchased and which carries out the fundamental housekeeping tasks needed to transfer data and look after the system generally.

PROGRAM: the set of instructions, put together by a programmer, which carry out a particular function such as the printing of a report, the updating of a file, playing a game, displaying results graphically.

PROTECTION: the operating system will protect records and files from overwriting by other programs or users.

PROM: see under ROM.

RAM: random access memory; this is the normal memory as found in every computer. It is called random access because the individual parts (called bytes or words) can be addressed in any sequence.

ROM: read-only memory; this is a form of memory which cannot have its contents destroyed by normal computer instructions; therefore it cannot be written to by normal computer instructions. PROM is the version of ROM which is "programmable" by a special device using a higher voltage. EPROM is the reusable form of PROM.

SECURITY COPIES: these are actual copies of precious programs or data files which are made by the user to act as a back-up in case of accidental erasure, spills of fruit juice, breakdown of the computer, loss of documents, etc.

SOFTWARE: the set of programs that operate the computer for a particular objective.

SYSTEMS SOFTWARE: the set of programs that carry out system housekeeping functions, often known as the OPERATING SYSTEM.

TAPES: a device for storing data and programs; cheaper than disk but slower and less versatile than disk; not particularly successful on micro systems.

TURNKEY SYSTEM: a set of programs and suitable hardware that requires minimum knowledge on the part of the user for successful operation; after the computer is switched on the programs prompt the user with suitable text to do the appropriate actions.

VDU or VISUAL DISPLAY UNIT: see MONITOR.

WORD PROCESSING: a set of programs that allow the user to type in text as if at a typewriter, store it magnetically, correct errors, amend and edit the text, add previously stored paragraphs, modify form letters, check the spelling, and many other functions; all without having to repeat previously correct work.

Book List

The following books are introductory texts which do not exclusively refer to one particular computer system but discuss basic principles. When a specific system is acquired it may be helpful to refer to a book which does illustrate the system acquired, in addition to the manufacturer's manual.

COMPUTERS AND COMMONSENSE
by R. HUNT & J. SHELLEY. 3rd edition pub PRENTICE HALL, 1983, Cost £5.95.

DEVELOPING MICROCOMPUTER-BASED BUSINESS SYSTEMS
by C. EDWARDS, published in association with ICMA (the Institute of Cost and Management Accountants).

POCKET GUIDE TO PROGRAMMING
by J. SHELLEY, pub PITMAN, 1981, £2.50.

POCKET GUIDE TO BASIC
by R. HUNT, pub PITMAN, 1981, £2.50.

OPERATING SYSTEMS POCKET GUIDE: INTRODUCTION TO OPERATING SYSTEMS
by L. BLACKBURN & M. TAYLOR, pub PITMAN, 1984, £2.50.

(There will be volumes on Unix, CP/M and MS-DOS in this series.)

A DICTIONARY OF COMPUTERS
by CHANDOS, GRAHAM and WILLIAMSON, pub PENGUIN REFERENCE BOOKS.

6

A Policy Statement on the Place of Computers and Information Technology in Schools in Britain

JOHN LEWIS

Malvern College, UK

Introduction

Schools in England have been fortunate. The Government believed sufficiently in the importance of computers to provide funds towards the cost of a microcomputer in every secondary and primary school in the country. Of course one microcomputer is not adequate in schools where many pupils want to make use of it and within a short while schools found the means to increase the number with the result that computer rooms with a large number of microcomputers are now commonplace in many parts of the country.

The computers have been used in many different ways—for courses in Computer Studies, to help in the teaching of other disciplines through the software which is becoming available, as a powerful tool in a science laboratory for recording and processing data, to enable students to learn how to write programmes themselves. Much imaginative work has been going on and new ideas are flooding the educational scene. One need, however, has been for a policy statement to identify what should be the place of computers within the curriculum. One of the first such statements was produced in June 1985 by the Independent Schools' Microelectronics Centre, which is based in Westminster College, Oxford. That statement is reproduced here, with the permission of the Centre, as an illustration of the thinking in at least one sector of English schools. It is hoped that it will be useful as a basis for discussion.

The Main Objective

The main reason for introducing computers into schools is to prepare all pupils for the inevitable contact they will have with them both in and out of school and later in their lives.

196

It is hoped that by the age of 16 all pupils will be able to use a computer and associated devices efficiently in a variety of tasks and across a range of subjects with particular attention being given to a problem-solving approach. It may well be that most pupils will own a computer by the time they are 16, but in any event all pupils should appreciate both the power and limitations of the computer and recognise its implications for society. The object is certainly *not* to turn them all into computer programmers.

Different Age Groups

In preparing this statement a distinction has been drawn between senior and junior schools. However, there is no particular age at which the abilities of pupils change suddenly so far as computing skills are concerned: what 15-year-olds in one school can do, 11-year-olds in another can achieve. For that reason we refer deliberately to "junior" and "senior" schools without being specific as to whether the age of transition is 11, 12 or 13.

Boys and Girls

There is plenty of evidence that girls are just as capable as boys when it comes to using computers (some would say better) and therefore no distinction is made between girls', boys' or mixed schools in this statement. However, there is some evidence that in *mixed* secondary school groups the girls may need more encouragement to persist in the face of social pressures.

The Aims in a Junior School

1. By the end of their time in a junior school, all pupils should be able to operate a microcomputer and to run a program with the same familiarity and relaxed confidence which they show towards other everyday technology.

    This will involve the use of a cassette or disk and handling the normal keyboard dialogue which is involved in the running of a program when responding to the instructions provided with the program.

2. All pupils should appreciate the ways in which computers can aid in the storage and retrieval of information.

    This will involve the use of a simple data-base related to themselves (e.g. a personal stamp collection) or their class (e.g. pupil data; names, addresses, birthdays, etc.) It could also involve the illustration of more sophisticated data-bases such as Prestel or local viewdata systems.

3. All pupils should have been involved in the use of a computer for simple word-processing.

    This will include the writing and correcting of short passages on the computer. Pupils should see how a computer can be used to improve pieces of their own writing, for example in a class news sheet. They can thus learn that a computer with a word-processing facility is a powerful aid in writing for a variety of purposes.

4. All pupils should appreciate that a computer can control other devices.

They should understand that a computer can control such devices as a printer, a washing-machine, a central heating system or a robot, and that it can respond to external devices such as a joystick, a temperature sensor, a clock/timer or lightmeter. In this work it would be good teaching practice to link the computer to the electronic devices which have been recommended for use in the science course in junior schools.

5. All pupils should have seen what is involved in programming in a simple language and should have had the opportunity of doing this for themselves.

They should have used a simple language with no more than ten words. (An example of such a language is used in the BBC program CARGO in which a gantry crane is used to load boxes on to a ship.) The simpler aspects of the language LOGO are particularly suitable for use in junior schools (and it has value at much more advanced levels too). It is not recommended that all pupils should be taught a more abstract and complex language such as BASIC, though there will be many pupils capable of mastering it and who will do so in their own time later. It would be sufficient if pupils were able to modify a given short programme of a few lines to effect something slightly different from that which it was originally doing.

It must be stressed that these are aims which are recommended for *all* pupils in a junior school. Many young people will be able to compile their own databases, do imaginative graphics work and write sophisticated programs. Schools should encourage this with such pupils, who will put the school computers to good use as well as their own computers at home.

The Aims in a Senior School

1. All pupils in a senior school should have experience, by the age of sixteen, of using a computer to run programs for a range of applications. As many as possible of these programs should relate to worthwhile applications in the school itself.

Examples might include a graphics package for presenting numerical information in graphical form (e.g. for the graphical interpretation of experimental data obtained by the pupil in the laboratory); 'business' type programs applied to the running of the school shop or to the materials store in the craft department. Other packages might include simulations of the use of computers in industry and commerce. These might include a spread sheet analysis package and a stock control package for the flow of goods through a warehouse.

2. All pupils should be able to assemble and organise data for entry into a computer data-base (there are many inexpensive ones available) which stores the information efficiently and then can be used to sort, order, re-order and retrieve it appropriately.

As well as building up data-bases for themselves, they should be shown the use of more complex and extensive data-bases of various kinds: for example census data (in history), weather data (in science or environmental studies), personal records (in school administration), reference material (in the library and elsewhere), travel bookings, etc. Methods of access should be shown, perhaps including Local Viewdata Systems with reference being made to Teletext, Prestel and Electronic Mail.

3. All pupils should be aware that a computer can be used as a word-processor and how it is possible to store and amend written material.

This should involve them handling the computer as a word-processor to produce their own material in a variety of contexts and/or subjects. They should know how to store, manipulate, correct and update such material and obtain a printed copy. Imaginatively used, a word-processor can enhance the craft of writing and improve the standards achieved.

4. All pupils should have some understanding of how a computer can control devices of various kinds.

This will require experience of a computer storing a sequence of instructions to control an external device through an interface. Some languages are particularly useful for control work, for example a sub-set of LOGO, often referred to as "turtle-graphics". Other languages or versions of languages, such as "Control Basic" have been written specially for control purposes. There will be other situations, for instance in the science laboratory, where the computer will be responding to the outputs from various sensors—temperature, light, radioactivity, pH, etc. This work should be closely linked to the electronics component of the science courses they do and to any technology work undertaken.

5. All pupils should appreciate the need to break down a problem into sections in order to structure a program and they should be able to write a simple program and to run it successfully.

This is commonly assumed to involve the learning of a well structured form of BASIC, but many other languages are equally appropriate. A small proportion of pupils may benefit from using the full range of commands and syntax of a sophisticated language and will need guidance on planning and structure.

Group Activities

Some of the most fruitful computer activities in schools are carried out in small groups working with one computer. For this reason a computer for every pupil in a class is not essential. The co-operative finding of solutions to problems in groups should certainly be included in any class.

Computer Examinations

Many schools will have pupils who are anxious to take a public examination in computer related studies. New syllabuses continue to appear. However, these are for the enthusiasts, rather than syllabuses which all pupils should be required to take.

It should be remembered that one more subject added to the curriculum means that pupils who chose it must drop one other optional subject. The proliferation of such courses can result in narrowing or imbalance in individual pupil's curricula.

Computer Assisted Learning (CAL)

In the same way that language laboratories and video tape recorders have contributed to work in the classroom, there is no doubt that computers have a great deal to offer in the teaching of many disciplines. More and more software is being developed, some of which is now of high quality. Good programs are available for use in the teaching of physics, chemistry, biology, mathematics, geography, economics and various languages. None of these is *essential* to good teaching, but teachers are well advised to become familiar with what is available and which programs can be used on the computers existing in their schools. They can take advantage of those programs which suit their courses and their style of teaching, and pupils will gain experience of another practical use of computers.

Conclusion

The impact of computers and the new information technology on communication makes it essential that schools keep abreast of developments. They will profoundly affect the curriculum and teaching.

7

Mastering of Microprocessor Technology

EDWARD W. PLOMAN

Vice Rector, Global Learning Division, United Nations University

THE Medium Term Perspective (1982–86) of the United Nations University (UNU) singles out two areas of advanced technology, biotechnology and microelectronics as being of major importance in current processes of industrial, economic and social transformation, both for industrialised and developing countries. Consultations concerning the potential role of the UNU in the areas of microelectronics and computer science led to the conclusion that a suitable focus would be the field of microprocessors. This paper summarises the context, as perceived by the University, and the general approach adopted for its activities in the field of microelectronics and gives an overview of principal activities.

Introduction

All over the globe, nations and their governments, international and inter-governmental organisations are suddenly awakening to the role information technology will play in future world growth.* Although the potential of information technology has existed since the development of the first computers in the early 1950s, its full realisation is only made possible today through the advent of the microprocessor and microcomputer. The pervasiveness and potential impact of current changes are expressed in the debate about the "information revolution" as reflected in different socio-cultural perspectives.

The microprocessor technology and its applications are becoming a key factor in the overall computer and information management fields.

---

* See the latest programmes of UN organisations (Unesco, ILO, UNIDO, UNCSTD, ACC Task Force on Science and Technology for Development, etc.), besides large-scale regional (EEC) and national schemes (France, Japan). Studies, books, conferences, appeals, resolutions abound at all the levels. The daily press and weeklies have plenty of stuff for their columns. Peripatetic experts are indefatigable in touring the globe.

201

Microprocessors embodied either in machines or computers are basically information-processing units which condition the control of products, processes and information flows. They represent an all-pervasive technology as increasing power is engraved on "chips" of decreasing size, greater capacity and falling cost. In light of the growing realisation of the seemingly illimited fields of application, the microprocessor appears as a new technological infrastructure in society with wide applications not only in industrial processes or in the up-grading of traditional technologies, but equally important, in all social activities which depend on the management of information, e.g. research and development (R&D) in science and technology, education, training and learning, public and private administration, communications and information services.

One of the main characteristics of microprocessors is their great dependence on scientific research and an advanced technical infrastructure. Currently, industrialised countries control this technology which is rapidly changing their economies, while at the same time reinforcing their traditional advantage in science and technology over developing countries. This is followed by an alteration of production systems also at the international level: these are becoming increasingly dependent on "information-intensive" sectors, concentrated in the rich countries, affecting at the same time the dominant forms and means of information flows. This situation is reflected in the growing importance of data bases and of accumulated expertise in computerised knowledge-based information systems in a few centres in the North.

At present, the basic components, i.e. the microchips, are produced only in a few countries, mainly in Japan and USA. Even though a number of other countries have decided to undertake research and development for the next generation of microchips (e.g. biochips), it is expected that the current situation will remain the same in the immediate future. This situation should also be seen in the context of the giant battle for industrial supremacy in both microelectronics generally and in the development of computer science: fifth generation computers, artificial intelligence and robotics.

Obviously, most countries will not be able to advance their own capacities in these areas in a manner which will make an impact. This fact, however, makes the microprocessor technology field that much more important, in view of basic differences between this technology and traditional large-scale main-frame computer technology. Traditional computer technology which in this perspective includes the coming fifth-general computers, etc. has been—and will be—available and transferred according to a "lock, stock and barrel" approach or to use another image, the "black-box" approach. In contrast, in microtechnology, it is possible to focus the understanding of technology on the underlying scientific principles and to move the mastering of technology to the level immediately following the production of the basic components. This in turn makes it possible to establish independent capacity in all areas, except the production of microchips, i.e. not only in the production of software but also in the production of hardware, i.e. microcomputers and associated equipment. This dimension may be symbolised by the "opening of the back-box" or seen as a dual-track hardware-software approach which is maintained in the UNU supported training activities involving young scientists from developing countries.

International Programmes

Due to the rapid advances in technology and the increasing impact of microelectronics in society, a number of organisations within the United Nations system are becoming increasingly involved in various aspects concerning the transfer of technology, the formation of national policies in developing countries and the analysis of the economic, social and cultural consequences in their respective areas of responsibility.

Diverse activities are undertaken by UNIDO (effects on industrial development), ILO (impact on employment), UNCSTD (transfer and use of micro-technology for development), while other UN agencies are implementing schemes for data processing (WMO, WHO, FAO) and communications development (ITU). The activities of Unesco in the field of information have largely focused on research and development, the establishment of information systems in developing countries and the use of informatics for socio-economic and cultural development. Other organisations involved in this field include IBI (Intergovernmental Bureau of Information), EEC (European Economic Community), OECD and other regional institutions.

Thus, many UN and other international organisations are becoming increasingly involved in various aspects of this new vast and rapidly evolving technological field. However, since each organisation is obliged to keep activities within its own area of responsibility, there is in this pervasive field no one international forum mandated to consider all inter-related aspects. In addition, research and development takes place outside of the international system which also has difficulties in attracting the best experts who work in private corporations or in universities and other scientific institutions. Needs and requirements in most developing countries are urgent and diverse, so that whatever is being done by international organisations can only satisfy a limited part of stated needs; the international programmes thus require full support everywhere also in order to counterbalance the preponderance of purely commercial interests. In this perspective, it seems crucial to have a clear view of accessible goals in a specific area and to endeavour to reach them by sustained activities over several years. Inconstancy, impatience and inconsistency are often reasons for the weakness and lacking impact of international programmes despite their alluring titles.

The UNU Programme

In the programme adopted by the University's Council for the biennium 1984–85 a cluster of activities is put under the general heading of "Information technology and society". For reasons mentioned in the introduction, a major area of activity focuses on microprocessor technology as the key element in current developments in which the UNU could provide a manageable and specific input in co-operation with institutions and experts in various regions of the world.

The statement in the UNU Charter that the University shall have as a central objective of its programmes the continuing growth of vigorous academic and scientific communities everywhere and particularly in the developing countries devoted to their vital needs in the fields of learning and research leads to two conclusions:

(i) the target groups should be at universities and research centres, particularly in developing countries; these are the direct, natural partners of the UNU in this area;

(ii) in the case of new and developing technologies where the R&D infrastructure of industrialised countries is so strong and large, the universities and other research centres in developing countries must play the role of initiators and hopefully innovators to satisfy vital needs, in close co-operation with the educational, industrial and service sectors.

The normal function of universities is for them to act as disseminators of knowledge, and especially in face of the paucity of other sources of reliable expertise in a country. Thus, the UNU must rely on universities and research centres, particularly in developing countries, as main actors of its projects in microelectronics research and training. Even, when the aim is to sensitise the governing structures in a country or to help the most indigent communities to benefit from the microprocessor based devices, the best way is *via* the local research and scientific communities. In addition, as centres of research and depositories of knowledge, the universities everywhere can act as unbiased partners in the transfer of technology. Through active investigation as well as through the channels of communication open for international exchange in the academic world, the universities can inject a measure of vigilance over the whole process of scientific and technological development in this area.

UNU programme activities in the area of microprocessor technology would normally be developed in terms of research, training and dissemination of knowledge. Since obviously the UNU would not be in a position to undertake hardware related R&D work in this field, priority has been given to training, primarily at the postgraduate level, coupled with support for specific research activities when and where required. These activities are complemented by the monitoring of advances in this field so as to guarantee a scientifically-sound, unbiased and up-to-date approach. In addition, research on the social and civilisational implications will be undertaken.

The dissemination of knowledge in this field encounters a different set of problems, due to the rapid pace of development, a certain degree of confusion caused by advertising and often loose speculations over the societal impact of new information technologies. For the UNU to attempt broad dissemination of information would be impracticable in this field which represents the fastest growing area of commercial publishing. The UNU would run the risk of doing no more than repeating known facts and duplicating the work of other institutions, including the media. Thus, in the area of microinformatics, it would be important to focus the dissemination activities on very precise objectives and clearly defined audiences, and provide close inter-linkages with other activities, in particular training and research, as well as the monitoring of developments and opportunities.

In general terms, the thrusts of the continuing programme in this area will be focused along three main lines:

— application of microprocessor technologies, particularly in developing countries and in selected fields which are of direct concern to the UNU, with a major emphasis on training and related research;

— monitoring advances in microelectronics (technology, applications, assessment of requirements);

— continued exploration of suitable research into the social and civilisational dimensions of the new information technologies based on microelectronics.

(i)   Application of Microprocessor Technology

The primary focus of activities concerning the applications of microprocessor technology will be on training, coupled with research when required and supported by the monitoring activities mentioned below, and, by assistance in up-grading laboratory facilities for training and research in or for developing countries.

The training encompasses all relevant hardware-software aspects of microprocessor technology and its applications for selected, specific purposes. The aim is to build up national capabilities, in Third World countries for mastering this technology. The concept of "mastering" is fundamental. As mentioned earlier, it makes all the difference when compared to usual "black-box" approach to technology transfer and "push-button" training for users of microcomputers and related devices.

The development of activities will take place along two selected axes:

(a) So far, the training and associated activities have concerned *microprocessor based systems and devices,* including microcomputers, both hardware and software aspects. The next phase for widening the areas of concern to the UNU lies in the field of *data base creation and management* on microcomputers, linked with the development of microcomputer networks, first as local area networks, later as long-haul networks. The state-of-the-art now makes it possible for smaller and "poorer" end-users to approach data-base management systems, so that even research in this area becomes meaningful. Applied to Third World environments, this would mean that developing countries could and should be recommended to set up local data bases, containing their own, locally produced data, and down-loaded data, selectively drawn from international and foreign data bases. This in turn means that unimpeded access, at home and abroad, must be ensured, through suitable physical structures (machines, programmes, networks) and through suitable legislation. A third aspect which requires further study for the identification of suitable UNU activities concerns software science and engineering *per se*.

(b) The second axis can be said to concern the substance of these applications. The UNU, through its co-operation with the International Centre for Theoretical Physics, is already engaged in what might be termed "microprocessors in the scientific laboratory". Activities in this area concern the organisation of regional training "colleges" in developing countries in 1984 and 1985, with the further objective of up-grading such courses in a second phase.

(c) The next steps comprise training and associated activities that will enhance developing countries' capacity in selected fields, according to stated needs and priorities in each case, or that are directly related to UNU activities and areas of concern. Thus, co-operation has been established with the Trinity College Dublin and the Irish authorities for the provision of special "tailor-made" training for young scientists in developing countries. Other activities include co-operation with selected African universities for projects comprising both training and support for microprocessor support units.

(ii) *Monitoring of Developments and Opportunities*

In this fast and inconsistently moving field, it is essential to continuously monitor advances in technology, to explore possible applications (in response to identified needs) and training opportunities. To this end, and in order to provide a sound basis for UNU programmes activities related to microprocessors technologies, it is proposed to set up a small monitoring group or alerting service on the broad and complex implications of the technologies.

Monitoring is usually conceived so as to include advice on operations to be undertaken and evaluation, based on interactive action. Such an activity would best correspond to the terms of reference of the UNU monitoring group, which would:

— review advances and trends in microprocessor technology with particular attention to applications, in Third World countries, in areas of the "societal infrastructure", e.g. education, research and development (R&D), public administration, communications;

— to assess the requirements and needs of end-users, particularly in developing countries, for information systems based on microprocessor technology;

— to advise the UNU on relevant programme activities and to periodically evaluate their implementation.

(iii) *Study of the Social and Civilisational Implications*

The study of the social and civilisational implications of microelectronics presents considerable difficulties. It must obviously be approached with clearly set objectives, be based on an adequate conceptual framework and be held to a manageable scope; studies in this area are being linked to analyses undertaken within the overall theme "Information Technology and Society", and current plans for development of activities in the area of new technologies in relation to employment and decentralised economic activities.

8

On the Use of Computers in Teaching in The Netherlands

HENK SCHENK

Laboratory for Crystallography, University of Amsterdam, Nieuwe Achtergracht 166, 1018 WV, The Netherlands

IN THE past years the computer has developed into a more and more useful aid for teaching. One of the most important reasons for this is undoubtedly the recent drop in prices and also the large scale availability of microcomputers. More essential, of course, are educational arguments in favour of the use of computers as a teaching aid; they are present as well. This paper deals with some aspects of the use of computers in teaching.

Applications of computers in daily life are manifold. It ranges from cars to washing machines, from salary-payment to word-processors, and from automatic pilots to distance teaching. Some knowledge of computers is therefore essential to everyone. However, not everyone has to know computers into the same depth. For comparison, everybody learns his mother-tongue in the first years of his life and continues until his death to renew and extend this knowledge. Those, who are able to profit from an educational system, learn also how to read and write their language. For most of them the level does not reach further than writing a letter to a friend or reading a simple book. Others manage the language better and they are developing into secretaries, teachers, copy writers, etc. However, only very few are becoming professional writers like novelists and poets, who master the language in all its aspects. Similar arguments apply to the level of understanding of computers and computer-applications. Few of us will be dealing with the design of computers and the system-programs necessary to run these computers. A few more people will be involved in the development of software such as compilers for computing languages, and other software-tools, which enables fast and easy production of applications. More people will be working on the applications itself; they produce programs which solve in a particular area a particular problem. It is these programs, which are used by again many more people to solve their particular problems. Finally, nearly everybody will push fingers on knobs of keyboards to input computers with signals, which drive them to fulfil specific

208

pre-programmed tasks. So therefore it is obvious that the computer deserves attention in teaching.

This rough description of the interaction between humans and computers is largely dependent on the availability of the latter. In the past 5 years a breakthrough is realised by the microcomputer. Computers with a power larger than 25-year-old main-frames are nowadays sold for a few hundred dollars and it can be expected that in the coming years computer power will continue to become cheaper. It is, of course, this change towards affordable prices which enables man to use computers on a wider scale, even in teaching.

How to use computers in teaching? In my view it is not necessary to teach pupils how to write a computer program, or how to realise applications, because only a few of them will develop themselves finally into programmers. We must concentrate on the majority that will just use computers plus existing software to help them solve their problems. What is then more natural for an educational system, than to use the computers to teach? In that case, by computer assisted learning, the students learn not only subject matter, but also learn to run the computer itself. In particular that subject-matter which is not absorbed easily by pupils can be chosen to be presented in this way.

The computer has a number of advantages as a teaching aid, of which I mention:

— the computer has an infinite patience with pupils;
— the computer is ideally suited for individual teaching (normally this is impossible in our teaching practice);
— by means of simulations new and challenging learning situations can be created;
— the computer is independent of the presence of a good human teacher and thus perfect for distance teaching (Open Universities are therefore very interested).

However, it is not a bed of roses; computers need software in order to teach, and in fact there is few useful teaching software public available throughout the world for any level of education. Nevertheless, scattered around the world individuals have written intelligent educational programs of high quality, but these are just used and known locally. Thus it is very difficult for individual teachers to search for, to find and to lay hands on sets of quality teaching programs suiting the needs of their pupils.

What is lacking, is a framework which guarantees the quality of teaching software, easy access to complete catalogues, and, sometimes, affordable prices. Maybe here is a task for Unesco, because certainly the fact that computer-driven teaching is independent of teachers, implies that when a good program is public available for a reasonable price it can be applied in any country and so it will help to spread out knowledge equally over the world.

9

Computers and the Future of Education

TOM STONIER

School of Science and Society, University of Bradford, UK

The Rise of the Knowledge Industry and the Economics of Education

Information is the most important single input into modern productive systems. It is no longer land, labour or capital—not even energy and raw materials. When you know enough—when you possess enough information—it is possible to reduce greatly the requirements for any of these inputs. For example, if land becomes expensive, as in the centre of a city, you build skyscrapers—when you know how to. When you know how to produce a tractor and such a tractor is more economic than a horse, then millions of acres around the world dedicated to growing fodder for horses become released.

Every time a robot displaces a worker, one has an example of knowledge displacing labour. The reduction in the requirements for capital, energy and materials is best exemplified by the computer industry itself. As the late Chris Evans was always fond of saying: "Had the automobile industry made as much progress in thirty years as the computer industry, it would be possible to buy a Rolls-Royce for £1.35, get three million miles to the gallon, and park six of them on a pinhead." Know-how, always know-how, advances modern productive systems.

The most valuable resource a country has is its **human capital**—the skills and education of its people. The matter shows up in countries with relatively small physical resource bases such as Switzerland, Japan, and even more dramatically, Singapore. These countries have little land, little mineral resources, and no oil. However, they have moved towards the top of the economic league and their citizens enjoy among the highest per capita incomes in the world. A great part of their economic success must be attributed to the high education levels.

It must be considered a truism that an educated workforce, including particularly its managers, learns how to exploit new technology whereas an ignorant one becomes its victim. This is why, in Western countries, some time

210

early in the next century, education will become the number one industry. At that time education will absorb a larger share of a country's GDP than any other economic activity—education in its various forms will become its largest single employer.

The gap between the West and most of the Third World countries is usually viewed in terms of material wealth. This is a gross over-simplification and probably misleading. Among the richest countries in the world is Kuwait. In terms of material wealth the oil rich countries have more than most Western countries. However, though they may be rich, they are still developing countries. The real gap between the West and the Third World is an information gap, or to be more precise, an education gap. That is where the real need is.

The Impending Revolution in Western Education

The first genuine revolution in education for over a century is beginning in some Western countries. It is seldom that a revolution begins **within** a well-established institution. The education system is no exception. That is not to say that there are not a large number of educators who would like to see the system change. However, the pressure will come from outside. It will be based on the emergence of cheap home computers, which will result in a significant shift from school-based back to home-based education.

This forecast is not derived from the assessment of educational needs or education technology, but simply from the economic dynamics of the situation. By January 1984, there were an estimated 2.1 million homes in the United Kingdom owning microcomputers. It is apparent that in the West the next consumer-boom will be in the area of home information technology.

At the moment, there is a shortage of good education software. This is the limiting factor in the spread of home computers. However, the situation will change rapidly during the mid-1980s as we move into the fourth generation of microcomputer education software packages.

The liking for computer games is virtually universal. The motivation for play is complex. It must be, at least to a large extent, biologically-based. Biologically useful activities activate the pleasure centres. A kitten chasing a ball, enjoys chasing as an activity in its own right. Considering the fact that the kitten will have to make its living as a hunter, chasing is a highly desirable activity. One of the questions computer psychologists need to answer is whether the almost compulsive preoccupation of boys with computer arcade games reflects their biological need to develop the hand-eye co-ordination required of human hunters who were not blessed with large claws or fangs, but did evolve hunting techniques using weapons.

Both our studies and those of other workers have observed that there are marked sex differences apparent in the enjoyment of certain games. The same can be said of profound differences in the response of children of various ages.

As an author of several educational programs, it becomes apparent that practice precedes theory.

Unfortunately, most adults still view the computer as a complicated piece of technology, well beyond their competence. Children do not have such inhibitions. To them the computer is a toy: it is associated with playing games. A four-year-old boy who had never used capital letters, when offered the "Letters Learn" program was thrilled by the game, yet had no interest in the alphabet books at home.

The Advantages of the Computer as a Pedagogic Tool

The computer without a doubt is the most powerful pedagogic tool invented since the development of human speech. The following lists some of the reasons for this:

1.   The most important reason is that the computer is interactive—unlike books, tapes, films, radio and television, the user's response determines what happens next. This gives children a sense of control. It also elicits active, motor involvement.

2.   Computers are fun. Human beings love to respond to challenges and they love to make things happen. The computer games industry has grown rich on that basic axiom. By coupling education to games of challenge, computer-assisted learning becomes fun.

3.   Computers have infinite patience. A computer does not care how slowly the user responds or how often a user makes mistakes. Among the earliest uses of computer-assisted learning on a wide scale, was the use of computers in Ontario in the late 1960s designed to help innumerate teenagers to fulfil the maths requirement for college courses. The programme was successful from a number of points of view, not the least of them the attitude expressed by one girl who stated that the computer was the first maths teacher who had never yelled at her.

4.   Good education programs never put a child down. Instead they provide effective positive reinforcement.

5.   Computers can provide privacy. Children, or for that matter teachers can make embarrassing mistakes without anyone seeing them. Ignorance, lack of skill, slowness to comprehend, poor co-ordination, all can be overcome in the privacy of one's home. The computer won't tell!

6.   At the same time, opposite to the fifth reason, the computer can be used in a variety of social situations. These include classroom activities involving groups, or a teacher and single pupil only, or two neighbourhood children, or party games, or a grandparent and grandchild, etc. Many education programs are designed to allow for either individual practice, or for two or more children to play games.

7.   The computers can explain concepts in a more interesting and understandable manner by means of animated material. No amount of talking,

writing, or providing diagrams, can compare with making things come alive on the screen.

8.   Whereas it is very difficult to hide things in a book, it becomes possible to hide things in a program which becomes apparent only on occasion. A book on re-reading holds few surprises (although the reader may have missed points the first time). In contrast, a computer program can be full of surprises. Good programs contain an element of mystery and uncertainty which keep the user interested. It means that the learning experience provides new situations not only for the students, but for the teacher or parent as well.

9.   The ability to simulate complex situations such as chemical reactions, ecosystems, demographic or economic changes is a particularly powerful reason for using computers in education. Training pilots, managers, doctors, chemical engineers, *i.e.* any profession or activity where a mistake in the real world could be very costly, is best served by learning on a computer which simulates the real-life situations. In addition, simulating real events often makes it possible to train students to think "laterally" across traditional subject boundaries.

The emergence of computer-based education will, over the next two decades have an impact as profound on Western society as did the establishment of mass education in the nineteenth century. For the Third World, as the new technology becomes cheaper, computers will allow them to catch up much faster. Those countries sufficiently educated to exploit the new information technology, could actually leap-frog over their Western counterparts.

10

Computers and the World of Large-scale Systems

DJORDJIJA B. PETKOVSKI

Faculty of Technical Sciences, V. Vlahovica 3, 21000 Novi Sad, Yugoslavia

1. Introduction

The current economic situation in the world, combined with increasing needs because of the growth in population and the level of civilisation, force most countries to give a hard look at the operating practices not only in their industries and agricultural sector, but in the whole society, i.e., in their socio-economic development.

Many have recently been led to the conclusion that something is seriously wrong with technology and their use of technology in the solution of complex, large scale problems facing society. In addition, modern technological systems often have a socio-economic interface which cannot be neglected, this includes, for example, economic interface with an energy system, psychological interface with a traffic system, etc. As such, socio-economic considerations have become important aspects of large scale systems. In this context, the nations have been especially concerned with problems including:

Energy resources
Food resources
Raw materials and waste rationalisation
Health care
Environmental pollution
Unemployment
Transportation, etc.

Needless to say, that all these problems and many others cannot be solved independently, and that many of them are considered to be "large-scale" by nature and not by choice. Although the notation of "large-scale" is a subjective one, the important point regarding large scale systems is that they often model real-life systems dealing with electrical power networks, energy systems, data networks, space structure, transportation, flexible manufacturing systems,

214

food, water resources, business, management, economy, environment, ecology, to name a few.

A number of researchers have paid a great deal of attention to the notation of "large-scale" systems, and many viewpoints have been presented on this issue. Although space and time are not available here to adequately discuss this issue, let it be mentioned that these systems are often separated geographically, and their treatment requires consideration of not only economic costs, as is common in "small-scale" centralised systems, but also such important issues as decentralisation, hierarchy, reliability of communication links, value of information, etc.

There is no precise, well-established definition of a large scale system. A system is usually defined as a set of resources of personnel, material, facilities and information which are organised to perform designated functions in order to achieve desired results. In this context, a possible definition is that large scale systems are those whose dimensions are so large that the classical, system techniques of modelling, analysis, controller design, optimisation, etc., fail to give reasonable solution with reasonable computational efforts.

Finally, it should be pointed out that large scale systems have received considerable attention because of their theoretical interest as well as practical importance.

2. Methodology and Technical Support

The real-life problems faced by many of us, and by many nations, have become enormously complicated. Take the examples of energy and food, or transport. To consider these problems independently of our social habits and psychological understanding is insufficient.

As the results of this, systems complexity and the complexity of individual system components (subsystems) have increased simultaneously. It can even be said that systems complexity increases with the square of technological progress. Moreover, the computational requirements for the system analysts and design increase in a polynominal progression with system complexity. Therefore, dealing with large-scale systems means coping with problems whose size increases several orders of magnitude for each order of magnitude advance in technology.

In other words, the modern systems engineers, planners, decision makers, etc., face new "problems of scale", and that is why the application of conventional techniques to large-scale systems with the aid of still larger and faster computers fail to give a reasonable solution.

Such large-scale systems can be treated from a number of viewpoints. One can focus on the information or decision problems facing the central or co-ordinating agents charged with controlling the aggregated behaviour of the system. Alternatively, one can focus on individual subsystems and their information and decision problems.

In this context, many researchers have paid a great deal of attention to various facts at large-scale systems such as modelling, model reduction, control, stability, controllability, observability, optimisation, and stabilisation, to name some of them. So far, there has not been developed a unifying complete theory to deal successfully with all these problems. However, many different, entirely new approaches to system analysis and design have been developed to cope with modern large-scale systems and their associated problems, that are so much a part of our society.

Another important point is that in large-scale system study, effective problem definition and problem solving require creative interaction among decision makers, planners, system analysts, computer specialists, mathematicians, and other specialists. In other words, strong, interdisciplinary research teams are needed in this case.

Computer science is today, and will be even more tomorrow, the key to industrial development, and therefore to economic and social change. The application of many theoretical results in large-scale systems have become meaningful only since the advent of digital computers. It should be pointed out that computers are the fundamental tool which makes it possible to deal with large-scale systems, in a far more comprehensive and systematic way. Computer technology has progressed so far that combinational interdisciplinary problems, arising in large-scale system theory and application, that took many hours to solve only a decade ago, can now be handled within a few minutes. Such technological advances are revolutionising the analysis, planning and design in large-scale systems. Until recently, for example, it was a time-consuming task to compute the technical, economic, financial and social implications of even a small number of variations of a given development possibility. These problems are now considerably overcome.

In this context, computer science should not be considered only as the knowledge of how to use a computer, i.e. it should not be considered as an additional "discipline" or "subject", but as a base for development of new possible methodology for other disciplines, including the theory and application of large-scale system.

Knowing that computer science will more and more influence various activities of society, and not wishing to fall into the trap of regarding the computer as an apparently magical tool—a panacea—we have to know exactly what we can do with a computer, the kind of problem it can solve, the questions that we can ask it, its constraints, its limits, and its possibilities. Introduction to computer science, moreover, is not an end in itself, and should be related to what is being taught at the same level in other disciplines, so that computer science takes on the character of an intersection subject. This is specially true in case of large-scale systems theory and application.

Finally, it should be pointed out that the computer itself is an equally large-scale system. Moreover, the computer network in which a number of computers are interconnected, become an essential inherent part of many large-scale systems.

3. Modelling

The first step in analysis, design and synthesis of real-life large-scale systems in the development of "mathematica model", which can be a substitute for the real problem.

A great deal of intellectual activity has been motivated by attempts to construct a "model" of a large-scale system. A model is an abstract generalisation of an object or a system. Any set of rules and relationships that describe something is a model of that thing.

It is essential that the underlying model is both realistic and correct; that it is easy to set and change parameters; that the running time is not too slow and that the results are presented in a clear and easily interpretable fashion.

A common use of a model is to provide the decision makers with a "laboratory" for exploring the alternative consequences of a wide range of alternative plans or strategies. In addition, a typical result of a modelling is the opportunity to see a system from several viewpoints, to polarise thinking and to pose sharp questions.

We have been using models in technical sciences for a long time. When we began to apply modelling to economics and the other social sciences, perhaps we were expecting too much. There are possibilities for effective use of mathematical models in large-scale systems, but there are limitations, and writing equations is not all there is to model-making. One person alone cannot make a valid model for large systems; the work requires an interdisciplinary team. Failed models are inevitably attributable to lack of working in teams.

One of the most important challenges to system theory brought forth by present-day technology, environmental, social and economic development is to overcome the increasing size and complexity of the relevant mathematical models. Since the amount of computational efforts required to analyse these processes grows much faster than the size of the corresponding system, the problems arising in large-scale systems may become either impossible or uneconomical to solve, even with modern computers.

It is common procedure in practice to work with mathematical models that are simpler, but less accurate than the best available model of a given system. In going from the most complex to the most simplified model the trade-off is between computational convenience and modelling adequacy. Also, not only should a model be faithful in terms of the physical reality that it represents, but it should also provide the planner or the analyst with enough information to enable him to act on the system in a knowledgeable way. In other words, a satisfactory model is a good aid to decision making which at the same time achieves the right level of trade-off between accuracy and computational convenience.

Therefore, in the examination of large-scale systems, the methods of analysis and design should somehow take advantage of special features in the system, leading to the simplification of the computing requirements. One important class of special features is the system structure. Lack of understanding of the structure of the underlying system can often lead to wrong conclusions regarding problem solution.

4. Simulation and Optimisation

Simulation, a system analysis approach, can provide a comprehensive view of a large-scale system. In case of development planning, for example, one can be interested in studying and approximating via a mathematical model, those relationships within the economy which are important in the development process. By translating this model into a computer language the likely final results of alternative development schemes can be evaluated through manipulation of the specified computer simulation model.

In essence, simulation is an iterative problem solving process which involves problem formulation, mathematical modelling, refinement and testing of the resulting model, and creative design and execution of simulation experiments intended to provide the answer to the question posed. This approach requires specialised knowledge from various professions and discipline, i.e. creative interaction among system analysts, computer specialists, decision makers, planners and other specialists.

In many real-life situations, decision makers are forced to make important decisions on the basis of a wide range of alternative plants and management strategies. In these cases, especially when large-scale systems are considered, the simulation approach can be time-consuming and not very efficient. Therefore, a systematic way is required to find the "best" solution. With current computing technology and methodologies, simulation models can also be used in an optimisation mode. This approach presupposes some criterion (index performance, welfare function), to be maximised or minimised. Therefore, it is usually necessary to decide upon a selection on the basis of which a choice can be made from among the many alternatives that are normally possible.

In other words, in this mode the model is programmed to automatically search for a solution which optimises the selected criterion, such as time, energy, profit, cost, etc. In addition, in many cases, an optimal policy is one that minimises or optimises the performance criterion without violating any chosen constraints. However, it should be pointed out, that optimal implies neither "good" nor "bad" but may simply be a way of selecting one policy from a number of satisfactory scenarios.

Another point is also very important. Once the optimisation problem is posed in a satisfactory way, its solution may be difficult because of the size, i.e. dimensionality of large-scale systems. Various simplifications of the optimisation methodology have been proposed. However, in the strictly mathematical sense, what is optimal in a simplified scheme, will not be, in general, an optimal solution to the original problem. Adequate experience is sadly lacking in this area and a great deal of further work is needed.

5. Decentralised Approach

There are systems that are composed of many interacting subsystems and are too large or too complicated in some sense to lend themselves conventionally to the various analytical or computational procedures for analysis and/or design that can be found in "classical" system theory literature. In other words, although many classically formulated, centralised techniques are theoretically applicable to systems of arbitrary size, a notable characteristic of most large-scale systems is that centrality fails to hold due to the lack of centralised information, centralised computing capability and/or numerical difficulties.

In addition, physically, large-scale interconnected systems composed of a large number of subsystems are usually large geographically and dimensionally, thus rendering the implementation of a classical centralised control law, i.e. decision making, with its required communication between all state variables of the subsystems.

As such, in a large-scale system one looks for decentralised controllers, i.e. decision making wherein only the states within geographically centralised subsystems need communication with another.

Therefore, for large-scale systems distributed widely in space, the scheme of decentralised strategy which is compatible with the information structure imposed by subsystems, has advantages in computation and implementation of control laws. In other words, the decentralised approach attempts to avoid difficulties in geographical separation, data gathering, storage, computer programme complexity, etc., which are compatible with the structure imposed by subsystems.

However, some communication among controllers, i.e. decision makers is not only realistic but also essential for successful operation of large-scale systems. The degree of communication affects each controller's behaviour to a large extent. The study of information structure concerns the specification of how much each controller (decision maker) knows about the system and how much of this he can use in his decision making.

The communication network for a decentralised large-scale system often takes the form of a computer network. As such, the efficiency and information transmissions of such a computer network may determine the degree of decentralisation required for a control system.

Another feature of a large-scale system is this hierarchical decision making aspect. For example, in socioeconomic systems, regional and national power grids, etc., optimisation is frequently to be carried out at several levels. For a two-level system, for example there are central or co-ordinating agents who are charged with controlling aggregate or average dynamic behaviour of the total system and several other local controllers.

6. Developing Countries

It should emphasise that the theory and applications of large-scale systems are rather new even for developed countries. And when we talk about the technological gap between industrialised and developing nations, the gap is particularly noticeable in the cases of large systems and computer technology. Overcoming this gap—trying to close it, is of particular importance.

Planning in any country, especially in developing countries, is a process fraught with uncertainty. Frequently, there are uncertainties about likely, immediate and long-range effects of development strategies. In addition, modern technological systems may have socioeconomic and environmental effects which cannot be neglected. The paucity of information, including computerised data bases, available for decision making is often cited in developed countries, with even more frequent mention in less developed countries. Poor communication facilities, especially in developing countries, often impede the accumulation of potentially available relevant information which might otherwise provide a reasonably well-informed basis for decision making.

Given these difficulties and uncertainties, a computer simulation model can have some conspicuous advantages for policy makers. It can serve as a means to experiment and to analyse comprehensively the complex relationships affecting the potential results of various policy alternatives.

However, substantial information reservoirs and research investigations are usually necessary in developing a reasonably complex and useful simulation model for regional or national development planning.

For many developing countries a trial and error, i.e. simulation approach would not be effective for the efficient utilisation of time, resources, etc., especially in the case of large-scale systems. This necessitates adopting a procedure that can effectively analyse and rank the alternative policies.

Methodologies of large-scale systems have been used very successfully over the last decade. However, the success of large-scale system methodologies in the solution of real-life problems in developed countries does not necessarily mean that the very same approach will succeed in solving the problem of a developing country.

For example, Vernon Ruttan raised an important question, referring to developing countries*: "Why is it relatively easy to identify a number of relatively successful small scale or pilot rural development projects but so difficult to find examples of successful rural development programs". In other words, for many developing countries additional efforts are needed to provide a base for successful solution of large-scale development problems.

* V.R. Ruttan, Integrated Rural Development Programs: A Skeptical Perspective, *International Development Review*, Vol. **XVIII**, No. 4, 1975.

The planning phase of large-scale system approaches for the solution of problems of a developing country naturally holds the highest priority. Each developing country, due to her unique social, economic, historical and technical assets and/or deficiencies, has got to plan her own strategy. The goals and needs of developing countries are very different from those of developed ones and also of each other. Therefore, any "expert" cannot function satisfactorily in any developing country if he is not very well versed in the background of this country. This background information cannot be acquired overnight. It is a slow process and, unlike expertise in a pure technical area, cannot be obtained by studying alone.

When technical support for large-scale systems analysis, planning and design is concerned, two remarks should be made:

(a) The computer programmes should be developed, if possible for widely available and popular personal computers. While this reduces computational efficiency somewhat, it will allow for the wide use of large-scale system methodology and tools by a broad range of professionals, institutions and enterprises, with minimum computer experience and minimum investment in providing computer equipment.

(b) It seems that a package transfer of existing methodology and technology with appropriate modifications to the real need of the country will be necessary, at least in the first phase of application.

If proper guidance could be made available in this field, modernisation, quality improvement, cost reduction, etc., could be expected in a broad range of large-scale systems.

The major benefits of modernisation in technical large-scale systems come from advanced control and process optimisation, not from personal reduction. For developing countries it is especially important that economic improvement is not achieved by personal reduction, because it makes a limited use of powerful equipment and only captures a small fraction of potential savings. Savings from better process control are in fact much larger than those achievable with personal reduction. In many studies the ratio was larger than 20 to 1.

Current limiting resources in building realistic large-scale models for developing countries appear to be:

(a) the amount and breadth of applicable descriptive information available about the interrelationship within the socioeconomic and environmental processes within the countries.

(b) trained personnel, i.e. the lack of expert knowledge to assist in the application of modern methodology for large-scale systems.

Trained personnel, in combination with access to appropriate computers must be available to build and continually update the functional relationships within the model and to interpret the results of model experimentation for the decision making.

Unfortunately, as known, the traditional school accepts very slowly, and to a limited extent, new methodology. The education system has to be more flexible to capture the latest trends and results in large-scale system theory and application.

In addition, interdisciplinary, joint research teams of experts from developing countries should be established. This includes scientific and technical information transfer between different countries. In other words, the exchange of knowledge should be regarded as an important part of international collaboration.

Corresponding institutions in developing countries should concentrate on the selected few research themes in large-scale systems which meet the real need of the people in the country, rather than attempt a whole range of possible research directions.

The efficient management and utilisation of part resources for a better world is a global problem. Therefore, everybody in every nation has something at stake in the successful design of such methodologies for the benefit of all.

7. Education

In the past, the education system put limitations on individual ability. To approach a complex problem from more than one point of view requires additional effort. These two facts had been constraints on the imagination. When large-scale system theory developed, it was considered to be a philosophical view—expressed by an over-use of mathematics. To develop a good theory, it is true that good mathematics and a general background in science help. Both theory and applications have progressed very much since the coming of "Big science".

New ideas and methods in large-scale theory are developed on a regular basis. These results should be examined on a continuing basis in order to present the principles of large-scale system theory and application to the students in the best manner possible. The syllabus should not be static and must be progressively modified in the light of experience and technological developments. In particular, it is important that the introductory courses in computers application to large-scale systems, for pedagogical reasons, include some practical work.

Examples presented to the students should be taken not only from mathematics (and other traditional sciences), but also from other fields according to the real needs of the country. In other words the problems have got to be chosen from life. Death or meaningless non-physical or unreal problems make the students become uninterested. The methodologies and design areas must excite the students so that their creativeness, innovative skills and virgin thinking capabilities can be brought out. In the beginning, the programming details should not be allowed to dominate the course.

The advent of the computer influences many academic subjects. Application of the computer to other disciplines where relevant should preferably be developed in the framework of those disciplines. In this context, it would also be desirable to develop the links between computer science and large-scale system theory and application. An introduction of computer studies and their interconnection with large-scale systems, will require careful reconstruction of the curriculum as a whole. In this connection a very important characteristic of this approach is its ability to create in the students an organisational algorithmic and operational attitude which is desirable for many lines of study.

The ability to simulate large-scale systems, is a particularly powerful reason for using a computer in education. Computers can explain concepts in a more interesting and understandable manner by means of animated material. In this context computer graphics will play an important role, as it makes things alive on the screen.

In addition, simulating real-life events, often makes it possible to train students to think "laterally" across traditional subject boundaries, which is especially critical in cases of large-scale systems theory and application.

Unfortunately, at the moment, there is a shortage of good educational software. This is the limiting factor in the spread of the large-scale system methodologies in the existing educational system. Therefore, one objective should be to develop a series of training aids in various forms for use in large-scale systems. In addition, there is an urgent need to exchange information on these and to make the material itself more suitable for school use and more accessible to teachers.

As mentioned, many of the large-scale system methodologies which have been developed to suit the needs of highly technologically developed countries, must be adopted to suit the requirements of the developing countries. It is self-evident therefore that developing countries with limited financial resources must be particularly careful to obtain the maximum benefit from their investment in development programs through the most effective utilisation and development of their staff.

Finally, it should be pointed out that, in general, education large-scale system studies should be a means not an end in themselves. They will help the students better to understand the world in which they will live.

8. Centre for Large-scale Control and Decision Systems

As mentioned, although a recent discipline, large-scale system theory has broadened its field of application from aerospace and other engineering problems to areas in which a mixture of economics, technological factors and human attitudes are an essential part of the problem. However, space and time are not available here to adequately discuss the issue of application. We will only mention some of the current activities undertaken at the Institute for

Measurement and Control, Faculty of Technical Sciences, University of Novi Sad.

The Institute for Measurement and Control is one of the departments of the Faculty of Technical Sciences, which was founded 25 years ago. During this period the activity was mainly concentrated on the process of education which produced more than 2,000 mechanical and electrical engineers. Every year we have over 300 new students. All of them have been educated in the field of control and system theory, and computer sciences, as well as in the fields of electronics, mechanics, physics, mathematics, measurement, etc.

In the field of control theory and computer sciences we give the following courses:

—Basic course in control theory
—Classical control theory
—Modern control theory
—Large-scale systems
—Introduction to optimisation methods
—Simulation and modelling
—Distributed parameter systems
—Stochastic system and estimation
—Programming of digital computers
—Basic principles of digital computers
—Logic design of digital computers
—Operating system theory
—Applied programming
—Information sciences and processes
—Computer networks
—Microprocessors and microcomputers
—Computer-based control systems
—Computer architecture and organisation

As known control and information theory and control systems design is a truly multidisciplinary area of education, research and applications. Typical application of large-scale control and decision system theory can be found in industrial control systems, chemical process control systems, robotics, control of nuclear reactors, aircraft, power systems, agricultural systems, economic systems, etc. etc. etc. Thus, at the Institute for Measurement and Control the *Centre for Large-scale Control and Decision Systems* was established as a centre for education, training and research in the area of Large-scale Systems Theory and Application especially in the following areas:

—Decentralised control systems
—Hierarchical control systems
—Computer networks
—High level languages for advance process control
—State space models for technical and economic systems

—Optimal, suboptimal and robust control systems
—Distributed parameter control systems
—Distributed computer control for networks of automatic processes
—Interactive computer graphics for simulation and design
—Automated process control
—Control for manipulators and robots
—Development of high level decision schemes, etc.

In addition to the courses for graduate level studies we offer other specialised programmes. These programmes include:

—Post-graduate Training Courses
—Doctoral Programme
—Advanced Programme in Applied Research
—Lecture Series
—Short Training Periods

Post-graduate Training Courses

Post-graduate Programme in computers, automisation and measurement is offered for students from Yugoslavia. This year we have nearly 50 students.

Doctoral Programme

An opportunity is provided to qualified candidates to participate in different programmes of research which can form the subject of a doctoral thesis for presentation at the University of Novi Sad.

Advanced Programme in Applied Research

Open to engineers who have a doctoral or master degree.

Leisure Series

One of two weeks specialised courses on specific topics of interest to engineers and scientists from industry, research institutes and university.

Short Training Periods

Qualified candidates are offered an opportunity to become acquainted with the work of large-scale systems through participation in a research project during a period of four weeks of more.

International Post-graduate Training Course in Large-scale
Control Systems

Under the sponsorship of Unesco, the Institute for Measurement and Control, Faculty of Technical Sciences, University of Novi Sad is organising an International Post-graduate Training Course in Large-scale Control Systems. The duration of the course is nine months (one academic year), and it is organised within the framework of the Unesco Programme of International Long-term Post-graduate Training Course in Science and Technology. This permanent international programme in research and training in the field of large-scale control systems, is the only one such a course in the field of large-scale systems sponsored by Unesco.

The course is directed largely, but not exclusively, to participants from developing countries. The course has as its subject Large-scale Control Systems: Theory and Applications, especially in their application in electric power systems, multivariable systems, economic systems, agricultural systems, etc. Its purpose is that of training the participants in order to acquire basic knowledge and knowledge about the practical application of the following branches of science: system theory, modelling and control of large-scale systems, optimisation, mini and microcomputers, informatics and computer networks. Particular emphasis is laid on the selection of a research field relevant to each applicant's home institute. The research training is provided through active participation in research groups.

The lecturers are the high level professors from leading Yugoslav universities and from leading universities throughout the world as well as experts from industry, research institutes and different international agencies and organisations. This year we have had lecturers from universities in Holland, England, USA and France, and experts from the World Bank.

This year, the interest for this course was far above our expectations. We have 31 participants from nearly 20 countries from Asia, Africa and Europe.

Appendix 2

Results of Questionnaire

A questionnaire was sent to all participants before the meeting and 80 replies were received. The number of ticks recorded in each box were as follows:

1. Teaching techniques and materials

(i) Please tick the appropriate boxes to show which of the following are in use in your locality at the levels indicated:

	Pre-school	Primary	Secondary	Tertiary	General Public
Slides	17	32	52	60	43
Transparencies for overhead projectors	9	20	46	63	35
Posters	39	56	58	44	53
Audio-tapes	21	28	39	39	34
Tape-and-slide packages	9	19	33	34	28
Experimental demonstrations	9	29	54	60	21
Films	19	40	57	55	43
Video-tapes	10	21	39	47	37
Video-discs	1	3	5	13	8
Microcomputers for simulating a system or an experiment		8	29	42	8
Microcomputers for data storage & retrieval		3	22	45	20
Microcomputers as interactive learning devices	3	16	27	39	8
Satellite television	1	4	6	7	14
Complete educational packages (i.e. booklets for teachers & pupils, slides, posters, etc. on a specific subject)	6	21	29	23	17

227

2. Transfer of Information to teachers

(i)   If an article or report on some aspect of science or technology teaching were to be distributed in your area (a) to all primary teachers, (b) to all secondary teachers, (c) to all teacher educators or (d) to all tertiary level teachers, how would it most likely be done?

(Please tick as appropriate)

Through the post	(a) 38	(b) 43	(c) 40	(d) 38
By a teacher organisation	(a) 19	(b) 26	(c) 16	(d) 10
By the Ministry of Education	(a) 47	(b) 48	(c) 27	(d) 17
Through a Teacher Training College	(a) 9	(b) 9	(c) 23	(d) 5
Through a University	(a) 3	(b) 3	(c) 20	(d) 42

(ii) Is the distribution of materials:

Done often	(a) 23	(b) 24	(c) 22	(d) 24
Done infrequently	(a) 34	(b) 37	(c) 33	(d) 31
Never done	(a) 6	(b) 5	(c) 7	(d) 9
Too expensive	(a) 5	(b) 5	(c) 4	(d) 2
Unnecessary	(a)	(b)	(c) 1	(d) 2

(iii) Are the regular methods of communication in your area (e.g. (a) telephone, (b) telex,(c) mail):
(Please tick as appropriate)

Dependable	(a) 36	(b) 30	(c) 50
Fairly dependable	(a) 24	(b) 8	(c) 18
Unreliable	(a) 7	(b) 4	(c)
Inexpensive	(a) 6	(b) 2	(c) 10
Too expensive	(a) 6	(b) 16	(c) 3

(iv) Are any of the following regularly arranged in your area?

Meetings for all teachers at a particular level (e.g. primary, secondary, etc.)	42
Meetings for all teachers in a particular subject (e.g. science, physics, technology, etc.)	56
In-service training courses for teachers in a particular subject	57
A local newsletter for all teachers at a particular level	28
A local newsletter for all teachers in a particular subject	31
A local newsletter for all in education	24

If so who arranges these (e.g. Science teacher association, Ministry of Education, etc.). Please give details:

Index